T0324143

Springer Theses

Recognizing Outstanding Ph.D. Research

Aims and Scope

The series "Springer Theses" brings together a selection of the very best Ph.D. theses from around the world and across the physical sciences. Nominated and endorsed by two recognized specialists, each published volume has been selected for its scientific excellence and the high impact of its contents for the pertinent field of research. For greater accessibility to non-specialists, the published versions include an extended introduction, as well as a foreword by the student's supervisor explaining the special relevance of the work for the field. As a whole, the series will provide a valuable resource both for newcomers to the research fields described, and for other scientists seeking detailed background information on special questions. Finally, it provides an accredited documentation of the valuable contributions made by today's younger generation of scientists.

Theses are accepted into the series by invited nomination only and must fulfill all of the following criteria

- They must be written in good English.
- The topic should fall within the confines of Chemistry, Physics, Earth Sciences, Engineering and related interdisciplinary fields such as Materials, Nanoscience, Chemical Engineering, Complex Systems and Biophysics.
- The work reported in the thesis must represent a significant scientific advance.
- If the thesis includes previously published material, permission to reproduce this must be gained from the respective copyright holder.
- They must have been examined and passed during the 12 months prior to nomination.
- Each thesis should include a foreword by the supervisor outlining the significance of its content.
- The theses should have a clearly defined structure including an introduction accessible to scientists not expert in that particular field.

More information about this series at http://www.springer.com/series/8790

Qingke Zhang

Investigations on Microstructure and Mechanical Properties of the Cu/Pb-free Solder Joint Interfaces

Doctoral Thesis accepted by
University of Chinese Academy of Sciences, Beijing, China

 Springer

Author
Dr. Qingke Zhang
Institute of Metal Research, Chinese
  Academy of Sciences
Shenyang
China

and

Zhengzhou Research Institute of Mechanical
  Engineering
Zhengzhou
China

Supervisor
Prof. Zhefeng Zhang
Institute of Metal Research, Chinese
  Academy of Sciences
Shenyang
China

ISSN 2190-5053   ISSN 2190-5061   (electronic)
Springer Theses
ISBN 978-3-662-48821-8   ISBN 978-3-662-48823-2   (eBook)
DOI 10.1007/978-3-662-48823-2

Library of Congress Control Number: 2015954609

Springer Heidelberg New York Dordrecht London

Printed on acid-free paper

Springer-Verlag GmbH Berlin Heidelberg is part of Springer Science+Business Media
(www.springer.com)

Parts of this thesis have been published in the following journal articles:

Zhang QK, Zhang ZF (2011) In-situ observations on creep-fatigue fracture behaviors of Sn-4Ag/Cu solder joints. Acta Mater 59:6017–6028 (Reproduced with Permission).

Zhang QK, Zhang ZF (2012) In-situ tensile creep behaviors of Sn-4Ag/Cu solder joints revealed by electron back-scatter diffraction. Scripta Mater 67:289–292.

Zhang QK, Tan J, Zhang ZF (2011) Fracture behaviors and strength of Cu_6Sn_5 intermetallic compounds by indentation testing. J Appl Phys 110:014502 (Reproduced with Permission).

Zhang QK, Zhang ZF (2012) Influences of reflow time and strain rate on interfacial fracture behaviors of Sn-4Ag/Cu solder joints. J Appl Phys 112:064508 (Reproduced with Permission).

Zhang QK, Zhu QS, Zou HF, Zhang ZF (2010) Fatigue fracture mechanisms of Cu/lead-free solders interfaces. Mater Sci Eng A 527:1367–1376 (Reproduced with Permission).

Zhang QK, Zhang ZF (2011) In situ observations on shear and creep-fatigue fracture behaviors of SnBi/Cu solder joints. Mater Sci Eng A 528:2686–2693 (Reproduced with Permission).

Zhang QK, Zhang ZF (2011) In-situ observations on fracture behaviors of Cu-Sn IMC layers induced by deformation of Cu substrates. Mater Sci Eng A 530:452–461 (Reproduced with Permission).

Zhang QK, Zhang ZF (2013) Thermal fatigue behaviors of Sn-4Ag/Cu solder joints at low strain amplitude. Mater Sci Eng A 580:374–384 (Reproduced with Permission).

Zhang QK, Zhang ZF (2009) Fracture mechanism and strength-influencing factors of Cu/Sn-4Ag solder joints aged for different times. J Alloys Compds 485:853–861.

Zhang QK, Long WM, Zhang ZF (2015) Growth behavior of intermetallic compounds at Sn-Ag/Cu joint interfaces revealed by 3D imaging. J Alloys Compds 646:405–411.

Zhang QK, Zou HF, Zhang ZF (2010) Improving tensile and fatigue properties of Sn-58Bi/Cu solder joints through alloying substrate. J Mater Res 25:303–314 (Reproduced with Permission).

Zhang QK, Zou HF, Zhang ZF (2009) Tensile and fatigue behaviors of aged Cu/Sn-4Ag solder joints. J Electron Mater 38:852–859 (Reproduced with Permission).

Zhang QK, Zou HF, Zhang ZF (2011) Influences of substrate alloying and reflow temperature on Bi segregation behaviors at SnBi/Cu interface. J Electron Mater 40:2320–2328.

Supervisor's Foreword

Soldering is the most widely used joining technology in microelectronic package, and the mechanical properties of the solder joints are important influencing factors on reliability of the microelectronic devices. With the increasing requirement on performance of the microelectronic device, the solder joints service in more and more severe environments; however, the understanding on damage behaviors of Pb-free solder joints is still lacking. Therefore, it is necessary to investigate the mechanical properties and damage mechanisms of the Pb-free solder joints for evaluating their reliability.

For this reason, in this study Dr. Zhang has designed a series of experiments to simulate the loadings suffered by the solder joints; the investigations include the fracture behavior of the interfacial IMC layers at the Cu/Pb-free solder interface, the tensile-compress fatigue damage behavior, creep-fatigue behavior, and thermal fatigue behavior of the Cu/Pb-free solder joints. Some innovative designs on mechanical property test method of the solder joints are applied, the damage behavior of the solder joints under different conditions are revealed, and the influences of interfacial microstructure, strain, stress, and temperature are comprehensively discussed. Overall, this work contains valuable information on reliability evaluation of the Pb-free solder joints.

Shenyang   Prof. Zhefeng Zhang
September 2015

Acknowledgments

I offer the sincerest gratitude to my supervisor Prof. Zhefeng Zhang. He gave me the chance to study in his research group. He has supported me throughout my work on mechanical properties and reliability of the Pb-free solder joints with his patience and knowledge while allowing me the room to work in my own way. One simply could not wish for a better supervisor.

I also owe my gratitude to Dr. Hefei Zou, it would not have been possible to complete my research without his help. Besides, my classmates and the fellow apprentices in the same research group have given me a lot of help.

I would like to acknowledge W. Gao, L.X. Zhang, J.C. Yuan, J. Tan, and Q.Q. Duan for the sample preparation, mechanical property tests, and microstructure observations during the experimental procedure. This work was financially supported by the National Basic Research Program of China under Grant Nos. 2010CB631006 and 2004CB619306.

Contents

Nomenclature

E	Tensile elastic modulus (Pa)
G	Shear modulus (Pa)
Q	Active energy (J)
R	Gas constant (J/mol K)
T	Temperature (K)
F	Force (N)
r	Radius (–)
b	Burgers vector (–)
S	Stress amplitude (MPa)
N_f	Fatigue life (–)

Greek Symbols

σ	Tensile stress (MPa)
ε	Tensile strain (–)
v	Poisson ratio (–)
τ	Shear stress (MPa)
π	Constant (–)
γ	Shear strain (–)

Chapter 1
Research Progress in Pb-Free Soldering

1.1 Soldering in Microelectronic Package

1.1.1 Microelectronic Package

The information technology based on electronic equipments has been developing very fast since 1960s, and has greatly promoted the progress of the industries and the improvement of human life. The electronic industry in China started late, but the government has paid much attention to its development, now China has become the world's largest production base of electronic products. The microelectronic industry will become more and more important in the economy.

The development of information technology relies on the progress of electronic equipments. The modern electronic equipments are usually very complex and consist of a large number of components. Only when the functions of the simple components are combined together through a few levels of package, they can transform into useable equipment. From a chip to a system, the package of electronic equipments can be divided into three levels as in Fig. 1.1. The modern packaging technology has become a complex system engineering involves package technology, material science, reliability assessment and so on.

Different technologies are used in different levels of electronic package. The connections in the first and second level package account for the majority of the connections, and the corresponding connection technology is known as the microjoining technology. For the microelectronic joining, the size effect should be considered because the components being jointed are very small, thus it is quite different from common welding technology. Its characteristics are as follow [1]:

(1) The components being jointed are small, thin, and light.
(2) The materials being jointed are nonferrous metals.
(3) The effects of surface tension, thickness of diffusion layer, dissolved quantity on the joining quality is significant.

© Springer-Verlag Berlin Heidelberg 2016
Q. Zhang, *Investigations on Microstructure and Mechanical Properties of the Cu/Pb-free Solder Joint Interfaces*, Springer Theses,
DOI 10.1007/978-3-662-48823-2_1

Fig. 1.1 Three levels of
electronic packaging

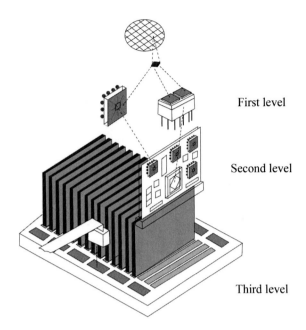

First level

Second level

Third level

(4) The requirement on joining accuracy is high.
(5) The joining process should have no effect on the electronic components.

With increasing circuit integration level, the number of joints in the electronic devices increases exponentially, while the size of the joints keep decreasing, thus the requirements of microelectronic joining technology become higher.

1.1.2 Soldering Technology

Among the microelectronic joining technologies, the most widely used technology is the precise soldering. Soldering is essentially a joining technology using the molten solder to react with the solid substrate material to form a thin reaction layer, and metallurgical connection is formed between the solder and the substrates after cooling [2, 3]. The metallurgical reaction of soldering used in electronic package is the same to common soldering, but its equipment and process are different from traditional soldering due to the size of the components being jointed. According to the heating method, there are iron soldering, impregnation soldering, wave soldering, reflow soldering, and so on. Among these technologies, the wave soldering is suitable for the traditional pin-through-hole (PTH) package, and the soldering process is shown in Fig. 1.2, while reflow soldering is used in surface mount technology (SMT) and ball grid array (BGA) package. The surface mount package and reflow soldering process are shown in Fig. 1.3.

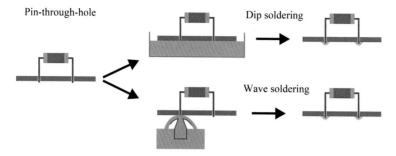

Fig. 1.2 Wave soldering and dip soldering in PTH package

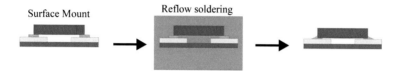

Fig. 1.3 Reflow soldering in surface mount package

1.2 Pb-Free Solders in Microelectronics

1.2.1 Sn–Pb Solder

The soldering technology has been used for over 5000 years. During its development process, the Sn–Pb alloy became the most popular solder because of its low melting point, superior wettability, and low price. The Pb element in the Sn–Pb solder can decrease the surface tension and improve the wettability, restrain the transformation of ductile β-Sn to brittle α-Sn, and significantly decrease the melting point [4, 5]. For a long time, the Sn–37Pb alloy is used as the solder in electronic package. However, Pb is a toxic element and has been included as one of the 17 most dangerous chemical substances to environment and human body. For this reason, major industrialized countries have legislated to forbid the use of the Pb-contained solders.

1.2.2 Requirements for Pb-Free Solders

To comply with the trend of Pb-free soldering, many countries have proposed their research and development plans of Pb-free solder. It is generally recognized that Pb-free solder should first meet the demand of environment protection and other toxic elements should not be added when Pb is eliminated. Besides, the soldering ability and reliability of the solder joints should be ensured, and the compatibility of

Pb-free solder with the current used devices and technologies should be considered. Specifically, the Pb-free solder for electronic package should fit the following requirements:

(1) The melting point of the Pb-free solder should be low, as close to the melting point of the Sn–37Pb eutectic alloy (183 °C) as possible.
(2) The wettability of Pb-free solder on the common used substrates or coating materials of the printed circuit board should be good.
(3) The thermal and electrical conductivity of the Pb-free solder need to be close to the Sn–37Pb solder.
(4) The strength, toughness, ductility, and creep resistance of the Pb-free solder should no less than that of the Sn–37Pb solder.
(5) The price of the Pb-free solder should be as low as possible.
(6) The Pb-free solder should compatible with the current used devices.
(7) The raw materials of the Pb-free solder should be easy to get.

However, it is not easy to find the Pb-free solder that can simultaneously satisfy the requirements above. In recent years, over 10 series of Pb-free solders have been proposed, but none of them can completely replace the Sn–37Pb solder.

1.2.3 Common Pb-Free Solders

After developed for over 20 years, more than 600 Pb-free solders have been proposed. The research progress show that the most possible substitutes of Sn–37Pb solder are the Sn-based alloys with the elements that can form low temperature eutectic, such as the Ag, Cu, Zn, Bi, and In. The compositions of common Pb-free solders are listed in Table 1.1 [4].

It can be found that the Pb-free solders are mainly Sn–Ag, Sn–Ag–Cu, Sn–Cu, Sn–Zn, Sn–Sb, Sn–Bi, and Sn–In alloy series. The microstructure and property of the main Pb-free solder series are as follows:

(1) Sn–Ag series: The Sn–Ag binary alloys are one of the earliest Pb-free solders. The Sn–Ag phase diagram is shown in Fig. 1.4 [6], which is a crystalline phase diagram, the eutectic temperature is 221 °C, and there is an Ag_3Sn intermetallic compound (IMC) region when the Ag content is about 75 %, the right side is similar to a binary eutectic phase diagram. The mutual solubility of Sn and Ag is very low, but they can form IMCs. At room temperature the Sn–3.5 wt%Ag eutectic structure is composed of β-Sn phase with little Ag content and thin Ag_3Sn particles disperse in the β-Sn [7]. The grain size depends on the cooling rate and the aging temperature. At the eutectic composition, the Sn–Ag alloy does not form eutectic structure directly, instead the β-Sn dendrite appears first, and then eutectic reaction occurs between the β-Sn. Microstructure of the Sn–3.5Ag at room temperature is shown in Fig. 1.5; the Ag_3Sn "particles" are actually needle-like in three-dimensional space. To

Table 1.1 Composition of common Pb-free solders

Solder alloy (wt%)	Sn	In	Zn	Ag	Bi	Sb	Cu
Bi–26In–17Sn	17	26			57		
Bi–32In		32			68		
Bi–41.7Sn–1.3Zn	41.7		1.3		57		
Bi–41Sn–1Ag	41			1	58		
Bi–42Sn	42				58		
Bi–43Sn (eutectic)	43				57		
Bi–45Sn–0.33Ag	45			0.33	54.7		
In–3Ag		97		3			
In–34Bi		66			34		
In–48Sn (eutectic)	48	52					
Sn–1Ag–1Sb	98			1		1	
Sn–1Ag–1Sb–1Zn	97		1	1		1	
Sn–2.5Ag–0.8Cu–0.5Sb	96.2			2.5		0.5	0.8
Sn–2.8Ag–20In	77.2	20		2.8			
Sn–25Ag–10Sb	65			25		10	
Sn–2Ag	98			2			
Sn–2Ag–0.8Cu–6Zn	91.2		6	2			0.8
Sn–2Ag–0.8Cu–8Zn	89.2		8	2			0.8
Sn–3.5Ag	96.5			3.5			
Sn–3.5Ag– <6Bi	90.5			3.5	6		
Sn–3.5Ag–1Zn	95.5		1	3.5			
Sn–3.5Ag–1Zn–0.5Cu	95		1	3.5			0.5
Sn–3.6Ag–1.5Cu	94.9			3.6			1.5
Sn–4.7Ag–1.7Cu	93.6			4.7			1.7
Sn–4Ag	96			4			
Sn–4Ag–7Sb	89			4		7	
Sn–4Ag–7Sb–1Zn	88		1	4		7	
Sn–10Bi–0.8Cu	89.2				10		0.8
Sn–10Bi–0.8Cu–1Zn	88.2		1		10		0.8
Sn–10Bi–5Sb	85				10	5	
Sn–10Bi–5Sb–1Zn	84		1		10	5	
Sn–4.8Bi–3.4Ag	91.8			3.4	4.8		
Sn–42Bi	58				42		
Sn–45Bi–3Sb	52				45	3	
Sn–45Bi–3Sb–1Zn	51		1		45	3	
Sn–56Bi–1Ag	43			1	56		
Sn–57Bi–1.3Zn	41.7		1.3		57		
Sn–5Bi–3.5Ag	91.5			3.5	5		
Sn–7.5Bi–2Ag–0.5Cu	90			2	7.5		0.5
Sn–0.75Cu	99.25						0.75

(continued)

Table 1.1 (continued)

Solder alloy (wt%)	Sn	In	Zn	Ag	Bi	Sb	Cu
Sn–0.7Cu (eutectic)	99.3						0.7
Sn–2Cu–0.8Sb–0.2Ag	97			0.2		0.8	2
Sn–3Cu	97						3
Sn–4Cu–0.5Ag	95.5			0.5			4
Sn–10In–1Ag–10.5Bi	78.5	10		1	10.5		
Sn–20In–2.8Ag	77.2	20		2.8			
Sn–42In	58	42					
Sn–5In–3.5Ag	91.5	5		3.5			
Sn–10In–1Ag–0.5Sb	88.5	10		1		0.5	
Sn–36In	64	36					
Sn–50In	50	50					
Sn–8.8In–7.6Zn	83.6	8.8	7.6				
Sn–5Sb	95					5	
Sn–5Sb	95					5	
Sn–4Sb–8Zn	88					4	
Sn–7Zn–10In ± 2Sb	81	10	7				2
Sn–8Zn–10In–2Bi	80	10	8		2		
Sn–8Zn–4In	88	4	8				
Sn–8Zn–5In–(0.1–0.5)Ag	86.5	5	8	0.5			
Sn–9Zn–10In	81	10	9				
Sn–5.5Zn ± 4.5In ± 3.5Bi	86.5	4.5	5.5		3.5		
Sn–6Zn–6Bi	88		6		6		
Sn–9Zn (eutectic)	91		9				
Sn–9Zn–5In	86	5	9				

optimize the Sn–Ag binary alloy, some other elements such as Cu, Bi, and Zn are usually added in it to form ternary or quaternary solder alloy.

(2) Sn–Cu series: The Sn–Cu alloy is similar to the Sn–Ag alloy, its binary phase diagram is shown in Fig. 1.6. In the phase diagram, there are a few IMCs at the Cu side, mainly Cu_6Sn_5 and Cu_3Sn. At low temperature and low Cu content, the Sn–Cu alloy can be recognized as binary eutectic alloy of Cu_6Sn_5 and β-Sn, the eutectic temperature is 227 °C, and the composition is Sn–0.7 wt%Cu. The Sn–Cu eutectic structure is similar to that of the Sn–Ag alloy, and also consists of β-Sn and the Cu_6Sn_5/Sn eutectic structure around the β-Sn, as in Fig. 1.7a [8]. Whereas, the Cu_6Sn_5 is not so stable as the Ag_3Sn. During the aging process, the Cu_6Sn_5 changes into plate-like Cu_6Sn_5 grain (see Fig. 1.7b), and results in decrease of tensile property and fatigue resistance of Sn–Cu solder alloy.

(3) Sn–Ag–Cu series: Through adding a little amount of Cu into the Sn–Ag solder alloy, the melting point of the solder can be further decreased without affecting the other properties, and the corrosion of solder to Cu can be decreased.

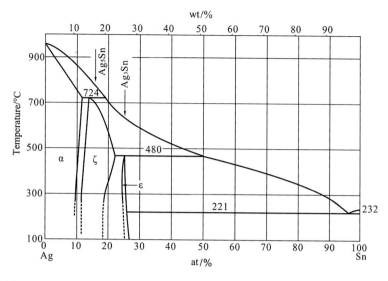

Fig. 1.4 Sn–Ag binary phase diagram

Fig. 1.5 Microstructure of Sn–3.5 wt%Ag eutectic alloy

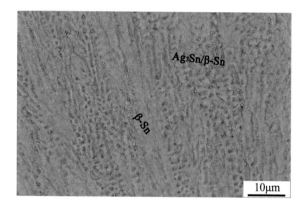

Besides, the mechanical properties of the Sn–Ag–Cu alloy are also higher than the Sn–Ag eutectic alloy. Because of its superior performance, the Sn–Ag–Cu alloy has become the standard Pb-free solder. The microstructure of the Sn–Ag–Cu alloy is also composed by the β-Sn and the eutectic structure, and the latter consists of Ag$_3$Sn and Cu$_6$Sn$_5$, as in Fig. 1.8 [9]. Since the solidification process of the Sn–Ag–Cu alloy is very complex, the exact eutectic composition has not yet confirmed; generally it is recognized that the eutectic point is around Sn–3.5wt%Ag–0.7 wt%Cu [9–11].

(4) Sn–Bi series: The Sn–Bi alloy is a widely used low temperature solder, its phase diagram is shown in Fig. 1.9 [6]. The Sn and Bi do not form IMC, thus the Sn–Bi alloy is pure eutectic structure. The most widely used Sn–Bi solder

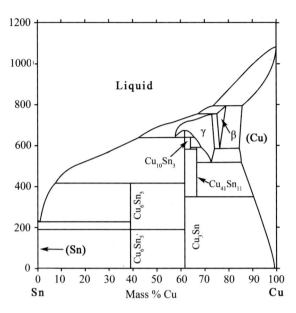

Fig. 1.6 Sn–Cu binary phase diagram

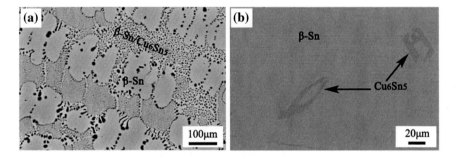

Fig. 1.7 Microstructure of Sn–0.7 wt%Cu eutectic alloy: **a** as-reflowed; **b** aged. Reprinted from Ref. [10], Copyright 2011, with permission from Elsevier

is the Sn–58 wt%Bi eutectic alloy, as in Fig. 1.10a, the Sn–58Bi solder shows typical eutectic structure. There are a few Bi elements dissolved in the Sn-rich phase, while the solubility of Sn element in the Bi-rich phase is very low. Since the Bi-rich phase is very brittle and its content in the eutectic structure is high, the shock resistance of the Sn–58Bi alloy is poor, but it still shows superior ductility at low strain rate [12, 13]. After thermal aging, coarsen of the Sn–Bi microstructure is obvious. Besides, embrittlement occurs at the long-term aged Sn–Bi/Cu interface due to the Bi segregation at the Cu₃Sn/Cu interface [14].

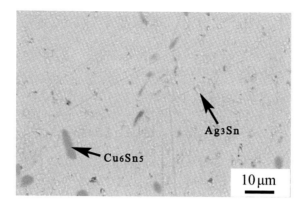

Fig. 1.8 Microstructure of Sn–3.8Ag–0.7Cu alloy

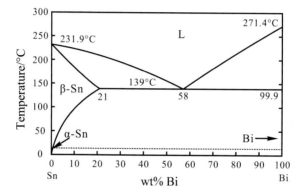

Fig. 1.9 Sn–Bi binary phase diagram

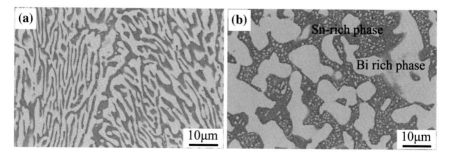

Fig. 1.10 Microstructure of Sn–58 wt%Bi eutectic alloy: **a** as-reflowed at 200 °C; **b** aged at 120 °C for 4 days

Fig. 1.11 Sn–Zn binary
phase diagram

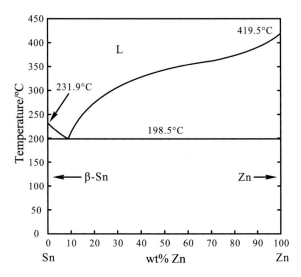

Fig. 1.12 Microstructure of
Sn–9 wt%Zn alloy. Reprinted
from Ref. [15], with kind
permission from Springer
Science+Business Media

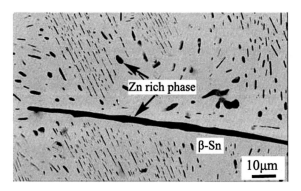

(5) Sn–Zn series: The melting point of the Sn–Zn alloy is very close to the Sn–Pb
solder, and the Sn–Zn solder shows superior mechanical property, makes it
has wide application prospect. The Sn–Zn binary phase diagram is shown in
Fig. 1.11 [6], it can be found that the Sn and Zn elements do not form IMC,
while the mutual solubility is very low. The eutectic composition of Sn–Zn
binary alloy is Sn–8.8 wt%Zn, and the eutectic temperature is 198.5 °C, only
15.5 °C higher than the Sn–37Pb solder. The eutectic structure is comprised of
β-Sn and a little Zn-rich phase, as in Fig. 1.12 [15].

1.3   Interfacial Reaction in the Solder Joints

Since both the Cu substrate and the Ni, Ag, Ag–Pd, and Au coatings can react with Sn and form IMCs, the solder joint interface in the circuit consist of three parts, i.e., the substrate, the solder, and the IMC between them. To form solder joints of high quality, the presence of IMC at the solders/substrate interface is usually desirable, while too thick IMC layers will degrade the reliability of the solder joints [2]. The volume ratio of the IMC layer is usually very high in the BGA solder joint, because the solder joint is small. With the continuing decrease in size of the solder joint, the volume ratio of the IMC layer will be higher. Therefore, the reaction behavior between the solder and the substrate or the coatings is very important influencing factors on properties of the solder joints.

During the soldering process, the substrate or the coating materials usually dissolve in the molten solder, and react with the Sn in the solder. The IMC grains nucleate at the solder/substrate interface, and the type of the IMCs depends on the composition of the substrate or the coating [4]. The formation of the IMC will consume the substrate element in the solder, makes more substrate material dissolve in the molten solder. The IMC grains will gradually cover the surface of the substrate and decrease the reacting rate. During the aging process, the IMC grains also grow slowly, and the shape of the grains changes obviously. The reaction behavior of Sn with common substrate materials or coating materials is as follows:

(1) Cu–Sn reaction: In the Cu–Sn phase diagram shown in Fig. 1.6, there are a few Cu–Sn IMCs in the Cu-rich region. When the temperature is lower than 350 °C, the Cu and Sn elements can form two IMCs, i.e., the Cu_6Sn_5 and Cu_3Sn [2, 4, 16, 17]. During the reflowing process, the Cu_6Sn_5 forms first, while the Cu_3Sn appears at the Cu_6Sn_5/Cu interface only after reflowed or aged for a certain time. Larsson et al. [18, 19] found that the Cu_6Sn_5 has two structures, the η' and η long period order superlattice structure, the former is stable below 350 °C, while the latter is stable when the temperature is over 350 °C. The Cu_3Sn IMC appears when the Cu/Sn atomic ratio is high, which also has a long period superlattice structure [2].

Fig. 1.13 shows the top views and cross sections of the as-reflowed and thermal aged Sn–4Ag/Cu interface. As in Fig. 1.13a, b, a large number of hemispherical or scallop IMC grains formed after reflowed at 260 °C for 3 min, which are Cu_6Sn_5 grains that nucleate at the interface and grow towards the solder. The IMC grains formed during the reflowing process is incompact, with wide gaps between them. After aged at 180 °C for 4 days, the thickness of the IMC layer grows obviously, and a Cu_3Sn layer appears at the Cu_6Sn_5/Cu interface. Besides, the hemispherical IMC grains transform into equiaxed grains, and the grain size is much larger, as in Fig. 1.13c, d.

(2) Sn–Ag reaction: In the Sn–Ag phase diagram, there are two IMCs, the first is disordered Ag_5Sn (ζ) phase, and the other is orthorhombic short-range order Ag_3Sn (ϵ) [2, 20–22]. During the reflowing process, the Ag also dissolves into

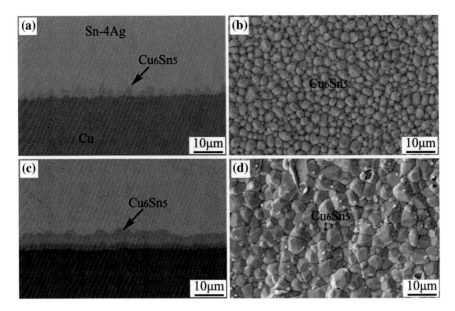

Fig. 1.13 Morphologies of the IMCs at the Sn–4Ag/Cu joint interface **a**, **b** as-reflowed at 260 °C for 3 min and **c**, **d** aged at 180 °C for 4 days after reflow

Fig. 1.14 Solidification structure of Sn–Ag–Pb (3.5 at.% Ag) solder showing the large Ag₃Sn flakes. Reprinted from Ref. [2], Copyright 2005, with permission from Elsevier

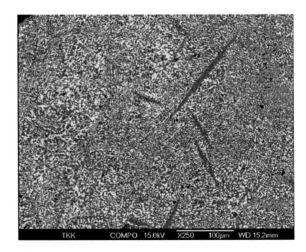

the Sn solder. After the Ag in the molten solder becomes oversaturated, the Ag$_3$Sn IMC separates out from the solder. When the Ag coating is thick, there is usually bulk Ag$_3$Sn formed in the solder, which is bad to mechanical property of the solder [23, 24]. When the Ag content is low, the Ag$_3$Sn usually forms thin particles or needles and improves the mechanical properties of the solder [2], while large Ag$_3$Sn flakes appear at the joint interface when the Sn solder reacts with the Ag substrate [25], as in Fig. 1.14.

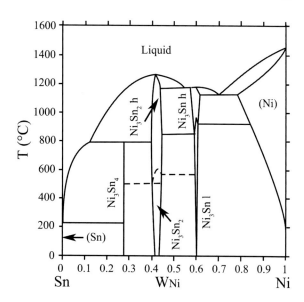

Fig. 1.15 Ni–Sn binary phase diagram

(3) Sn–Ni reaction: The Ni element can also form a few kinds of IMCs with the Sn element [26], but the reaction rate is very low compared with the Sn–Cu reaction, and the thickness of the IMC at the Sn–Ni interface is far lower than that at the Sn–Cu interface. Therefore, the Ni can be used as coating to restrain the growth of the IMC layer. The Ni–Sn binary phase diagram is shown in Fig. 1.15. As in the figure, when the reflow temperature is lower than 260 °C, the Ni_3Sn_4 appears first at the Ni/Sn solder interface [27–29]. With increasing reflowing or aging time, usually there is still no Ni_3Sn_2 or Ni_3Sn appearance at the interface. Besides, the Ni_3Sn_4 layer formed during the reflowing process does not show a parabolic grow, indicating that growth of the Ni_3Sn_4 IMC may not be controlled by diffusion mechanism [30]. With increasing reflowing time, three different forms of Ni_3Sn_4 IMC grains appear at the Ni/Sn interface [31].

In general, the appearance of IMC between the solder and substrate is a signal of good metallurgical connection, while too thick IMC layer will decease the reliability of the solder joint [2], because the IMC phase is usually brittle, Therefore, restraining the growth of the IMC layer through alloying the solder or control the soldering parameters is an important research direction.

1.4   Reliability of Solder Joints in Electronic Circuit

1.4.1   Loadings Suffered by Solder Joints

Since the solder joints in the electronic device service in complex thermal, mechanical, and electronic conditions, the joints may failed in many different modes [32]. In fact, over 70 % of the breakdown of the electronic device is induced by failure of the solder joints. The BGA solder joint is the most common solder joint in the microelectronic device; its shape and location are shown in Fig. 1.16. In the electronic components the primary strain applied on the solder joints results from the differences in the coefficients of thermal expansion (CTE) between the chip, chip carrier, and circuit board. Since the electronic equipments in service are periodically turned on and off, cyclic loading occurs in the solder joints and fatigue damage is induced. Besides, for the unfixed electronic equipments, vibration, move or impact can also result in loadings applied on the solder joints. For the portable electronic products, sometimes the shock loading is serious [4]. Since the melting points of the Sn-based solder are low, the solder can creep even at room temperature, and high temperature recovery and recrystallization occur when the strain is high [4, 33, 34], making the deformation mechanism of the solder joint more complex. The major mechanical loadings suffered by the solder joints are as follows:

(1) Shear loading: The shear loading is the primary loading suffered by the solder joint. Because the CTEs between the substrate and the chip are quite different, when the Printed Circuit Board (PCB) is heated, the difference in thermal deformation of the chip and the substrate will result in a shear strain on the solder joint [35], as in Fig. 1.17. The shear strain depends on the distance between the solder joints, the size of solder joints, the difference in CTE, and the temperature.

(2) Tensile loading: When the service temperature is high, thermal flexure may occur in the polymer substrate and makes the solder joints at the edge of the BGA suffer tensile strain, as in Fig. 1.18. Similar phenomenon may also appear when the electronic products are transported, or when they are tested in fixed equipment [4]. Because the solder joint is small, a little tensile deformation will result in a high tensile strain.

(3) Creep loading: When an alloy service in a temperature higher than 50 % of its melting temperature and the loading is high (but far lower than the yield

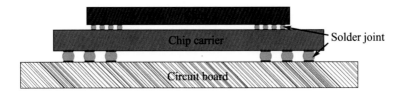

Fig. 1.16 BGA solder joints in microelectronic devices

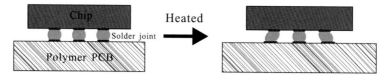

Fig. 1.17 Shear deformation of the solder joints when the microelectronic device is heated

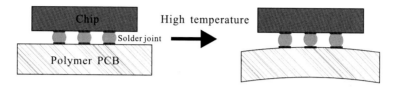

Fig. 1.18 Tensile deformation of the solder joints when the board is subjected to thermal bending

strength), creep deformation becomes a significant deformation within a long period. Because the melting points of the Sn-based solders are very low, their homologous temperature (T/T_m) are usually higher than 0.6 even at room temperature. Therefore, the creep deformation is significant. At low strain rate, creep deformation is no less than common plastic deformation [36].

(4) Fatigue loading: Since the electronic equipments in service are periodically turned on and off, some local circuits switch frequently, cyclic thermal deformation occurs in the electronic and the solder joints will suffer cyclic loading, i.e., fatigue loadings. Besides, the thermal strain rate is usually very low and makes the solder can easily creep, thus deformation of the solder is actually creep-fatigue deformation.

1.4.2   Mechanical Properties of Pb-Free Solders

The mechanical property of the Pb-free solder is one of the most important factors for selecting the solder. To make a comparison, there have been many investigations on mechanical properties of the solders, mainly focus on the tensile strength, shear strength, creep resistance, and the fatigue property.

(1) Tensile and shear strength: The yield strength, tensile strength and elongation of the Sn–37Pb solder and some common Pb-free solders are given in Table 1.2 [37–41]. As in the table, generally the mechanical property of the Pb-free solder is good, their tensile strength are higher than the Sn–37Pb solder. At different strain rates, the difference in tensile strength of a solder is

Table 1.2 Tensile properties of Pb-free solders

Solder	Elastic modulus (GPa)	Yield strength (MPa)	Tensile strength (MPa)	Elongation (%)
Sn–37Pb	39 [37]	–	19 (20 °C), 4 (100 °C) [38]	–
Sn–58Bi	42 [37]	41 [37]	–	20 (20 °C), 159 (85 °C) [39]
				73–150(0.0033–0.005) [40]
Sn–58Bi–1Pb	–	–	73 (0.4/s, cast) [37]	–
Sn–3.5Ag	50 [37]	48 [37]	42.8 (cooling rate 5 °C/min) [41]	42.5 (cooling rate 5 °C/min) [41]
			55 (0.022/s, cast) [37]	
			37 (3.3×10^{-5}/s, cast) [37]	
			20 (1.5×10^{-4}/s, cast, 25 °C aging) [37]	
			56 (8×10^{-4}/s, cold drawn) 37 (25 °C) [38]	
Sn–3.5Ag– <6Bi	42(1Bi) [28]	43(1Bi) [28]	71.7 (cooling rate 5 °C/min) [41]	15 (cooling rate 5 °C/min) [41]
				40 (1.5×10^{-4}/s) [40]
				25 (0.033) [40]
				31 (3.3×10^{-5}/s) [40]
Sn–3.3Ag–4.7Bi	–	75 (6.56×10^{-4}/s, 27 °C) [33]	82 (6.56×10^{-4}/s, 27 °C) [28]	10 (6.56×10^{-4}/s, 27 °C) [33]
Sn–3.5Ag–1Zn	–	–	52.2 [41]	27.5 [41]
Sn–0.7Cu	–	22 (6.56×10^{-4}/s, 27 °C) [33]	23 (6.56×10^{-4}/s, 27 °C) [33]	45 (6.56×10^{-4}/s, 27 °C) [33]
Sn–3.1Ag–1.5Cu	–	45 (6.56×10^{-4}/s, 27 °C) [33]	48(6.56×10^{-4}/s, 27 °C) [33]	36(6.56×10^{-4}/s, 27 °C) [33]
Sn–3.5Ag–1Zn–0.5Cu	–	–	48.3 [41]	7 [41]
Sn–9Zn		–	64.8 [41]	45 [41]

[a]It is the strain rate in the brackets

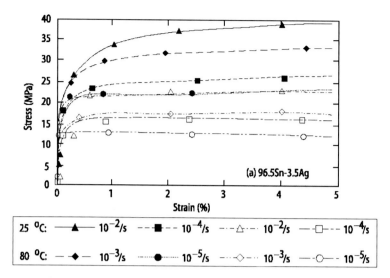

Fig. 1.19 Tensile curves of Sn–3.5Ag solder at different temperatures and strain rates. Reprinted from Ref. [46], with kind permission from Springer Science+Business Media

obvious. Investigation on strength of the solder has proved that the relationship between the tensile strength and the strain rate is as follows [42–45]:

$$\sigma = C\dot{\varepsilon}^m \tag{1.1}$$

where σ is the tensile strength, $\dot{\varepsilon}$ is the strain rate, m is strain rate sensitivity coefficient and C is a constant.

Due to the low melting point, the temperature is an important external influencing factor on tensile strength, as in Fig. 1.19 [46]. That is because the creep and superplastic deformation are easier to occur in the solder at high temperature and low strain rate and makes the strength decrease and the elongation increase. The solidification rate and thermal aging can affect the yield strength of the solder through influencing its microstructure [47, 48].

The shear loading is the main loading suffered by the solder joint, and the shear property of the solder and solder joint are important influencing factors on reliability of the solder joint. The relationship between shear modulus and tensile modulus of the metals is as follows:

$$G = \frac{E}{2(1+v)} \tag{1.2}$$

where G is the shear modulus, E is the tensile elastic modulus, v is the Poisson ratio (~ 0.33). The shear properties of common Pb-free solder are shown in Table 1.3 [37, 40, 49–52].

Table 1.3 Shear properties of Pb-free solders

Solder	Shear modulus (GPa)	Shear strength (MPa)
Sn–40Pb	–	34 (20 °C), 21 (100 °C) [49]
Sn–58Bi	15.8 [4]	35 (0.0033–0.005/s) [34]
		55 (0.4/s) [50]
		23.7 (0.4/s) [51]
		26 (0.001/s) [40]
		26 (0.00062/s, 25 °C) [37]
		28 (0.0015/s, 60 °C) [37]
		9 (0.004/s, 100 °C) [37]
Sn–3.5Ag	18.8	38 (20 °C), 23 (100 °C) [49], 27 (0.004/s, 25 °C) [37]
		39 (4/s, 25 °C) [37], 14 (0.004/s) [48], 55 [52]
Sn–3.5Ag–5Bi	–	35(0.0033–0.005/s) [37]
Sn–5Sb	–	37(20 °C), 21(20 °C) [49]

[a]It is the strain rate in the brackets

(2) Creep behavior: Obvious creep deformation occurs in the solder at room temperature and low stress. It has been well accepted that dislocation climb and grain boundary sliding are two major creep deformation mechanisms for the Pb-free solders, the creep deformation is the sum of the contributions of the two mechanisms, and their contributions depend on the strain rate, stress level, and temperature [33].

Similar to typical creep curves of metallic materials, the low cycle creep-fatigue process can also be divided into three stages according to the increasing rate of strain, as in Fig. 1.20 [53]. The second stage is the most important because it occupies most of the creep process and the increase rate of strain is simple [54]. The most usually used equation to describe the deformation behavior of the second stage is the Dorn power law equation [55]:

$$\dot{\varepsilon} = A\sigma^n \exp\left(\frac{-Q}{RT}\right) \tag{1.3}$$

and the Garofalo hyperbolic sine equation [56]:

$$\dot{\varepsilon} = C[\sin h(\alpha\sigma)]^n \exp\left(\frac{-Q}{RT}\right) \tag{1.4}$$

where $\dot{\varepsilon}$ is the creep strain rate, σ is the tensile stress, n is the stress factor, and Q is the active energy. Q and n depend on the main creep mechanism, R is the

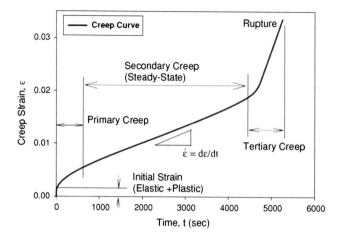

Fig. 1.20 Typical creep curve for Pb-free solder. Reprinted from ref. [53], with kind permission from Springer Science + Business Media

gas constant, T is the temperature, A and C are constants related to the structure of the solder. Thus far many investigations on creep behavior of the solder aim to build up the constitutive equations of the Pb-free solder at different conditions, i.e., to get the stress factor and active energy of the solder [57–61]. Through studies on constant creep rate of solder, the creep parameters of different series of solders has been obtain, as in Tables 1.4 and 1.5 [62–66]. However, the creep deformation of solder is complex, creep behavior of the solders needs to be further investigated.

Through adding some nano-sized materials in the solder, the creep resistance can be improved, because the particles can restrain the dislocation movement.

Table 1.4 Creep parameters of Garofalo hyperbolic creep models of solder alloys

Solder	Creep parameters				Test method and reference
	C	α (MPa^{-1})	n	Q (KJ/mol)	
Sn–37Pb	0.158	0.406	1.38	50.0	Cast bulk, tensile [62]
	10	0.1	2	44.9	BGA solder joint, tensile [63]
Sn–3.5Ag	178.5	0.115	4.75	57.1	Bulk, tensile [63]
	23.17	0.0509	5.04	41.6	Bulk, tensile [64]
Sn–3.0Ag–0.5Cu	2631	0.0453	5.0	52.4	[65]
Sn–3.9Ag–0.6Cu	0.184	0.221	2.89	62.0	Cast bulk, tensile [62]
Sn–3.8Ag–0.7Cu	32,000	0.037	5.1	65.3	Bulk, tensile [66]

Table 1.5 Creep parameters of Dorn power law creep models of solder alloys

Solder	Creep parameters			Test method and reference
	A (s^{-1})	n	Q	
Sn–3.5Ag	5×10^{-6}	11	79.8	BGA solder joint, tensile [63]
	9.44×10^{-5}	6.05	61.1	Bulk, tensile [64]
Sn–4.0Ag–0.5Cu	2×10^{-21}	18	83.1	BGA solder joint, tensile [63]

Investigations of Shi, Guo and Tai [67–70] reveal that the creep resistance of solder can be improved through adding Ni, Ag, and RE nano-sized particles.

(3) Creep property of solder: Because the solder joints in the microelectronic device service in dynamic thermal-mechanical loadings, the effects of fatigue property on reliability of the solder joints is significant, while the data on fatigue property of the solder is lacked, and the test condition in different studies is quite different [71–78]. It is generally recognized that the fatigue property of the Pb-free solder is higher than the Sn–37Pb solder [4, 79, 80]. Lea [80] report that the fatigue property of the Pb-free solder increases according to the following order: 64Sn–36In, 42Sn–58Bi, 50Sn–50In, 99.25Sn–0.75Cu, 100Sn, 96Sn–4Ag, 95Sn–5Sb. Kanchanomai, Pang, and Shang find that the fatigue lives of Sn–Ag, Sn–Cu, and Sn–Ag–Cu solders are affected by the temperature and loading frequency [74–78], as in Fig. 1.21, the fatigue life decreases obviously when the temperature and the frequency decrease. The fatigue crack initiation and propagation behavior of the solder is also related to the frequency. The surface fatigue crack initiation morphology of the Sn–3.8Ag–0.7Cu solder is shown in Fig. 1.22, it was found that the fatigue cracks appear at the grain boundaries. The heat treatment can also affect the fatigue property of the solder through affecting the microstructure.

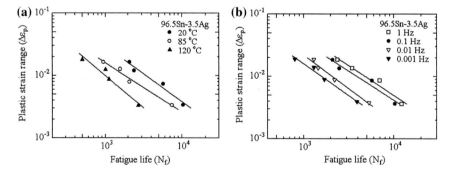

Fig. 1.21 Plastic strain range–fatigue life relationships of Sn–3.5Ag solder under various **a** temperatures and **b** frequencies. Reprinted from Ref. [75], with kind permission from Springer Science+Business Media

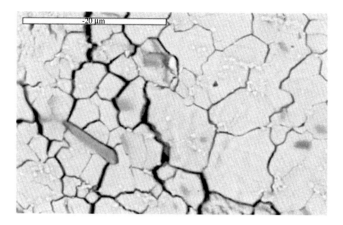

Fig. 1.22 Fatigue crack initiation in equiaxed Sn–3.8Ag–0.7Cu alloy. Reprinted from Ref. [78], with kind permission from Springer Science+Business Media

1.4.3   Fracture Behavior and Strength of Solder Joints

The strength of the solder joint is essential in evaluation of the solder joint which depends not only on the solder but also on whether a good joint interface is formed. As discussed before, the Sn-based solder can form different IMCs with the substrate or coating materials, and the thickness of the IMC layers increases with increasing aging time, makes the fracture location of the solder joint transform from the solder to the IMC layers, and the strength decreases obviously [81–84]. The decrease in tensile strength of the Sn–3.5Ag/Cu solder joint during the aging process is shown in Fig. 1.23 [82], in which the strength keeps decreasing with increasing aging time. The corresponding fracture surface is shown in Fig. 1.24 [82]. When the aging time is short, fracture occurs in the solder close to the joint interface, and the dimples appear at the fracture surface; after aged for a long time, the solder joint fracture inside the IMC layer in a cleavage mode [82]. With increasing thickness, the inner stress will accumulate in the IMC layer [85, 86], making the strength to decrease.

The evolution in shear strength of the solder joint during the aging process is similar to that of the tensile strength, while the fracture behavior is different [87–94]. Investigations of Deng and Yoon et al. [87, 89] reveal that the solder joint also fractures in the solder close to the joint interface under shear loading, only the size of the dimple in the fracture surface changes, as in Fig. 1.25. Although the shear fracture behavior of the solder joints are affected by the thickness and morphology of the interfacial IMCs, shear strength of the solder joints dependent primary on the solder [87]. The coarsen of the solder during the aging process makes the strength of the solder decrease, further result in decrease in strength of the solder joint and transform of the fracture mechanisms.

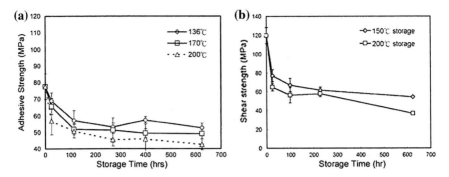

Fig. 1.23 Decrease in **a** tensile strength of the Sn–3.5Ag/Cu and **b** shear strength of Sn–4.11Ag–1.86Sb/Cu solder joints during the aging at different temperatures. Reprinted from Ref. [82], Copyright 2003, with permission from Elsevier

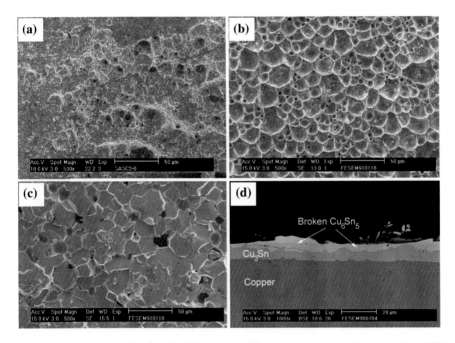

Fig. 1.24 Fracture morphology of solder joint subjected to tensile loading when the IMC thickness is **a** less than 1 μm; **b** between 1–10 μm; **c** thicker than 10 μm; **d** side surface. Reprinted from Ref. [82], Copyright 2003, with permission from Elsevier

The strength and fracture behavior of the solder joint depend not only on the strength of the solder and the IMC thickness, but also affected by the strain rate. At high strain rate, the solder joints are more apt to fracture in the IMC layer, and the strength is higher than that at low strain rate [81, 95].

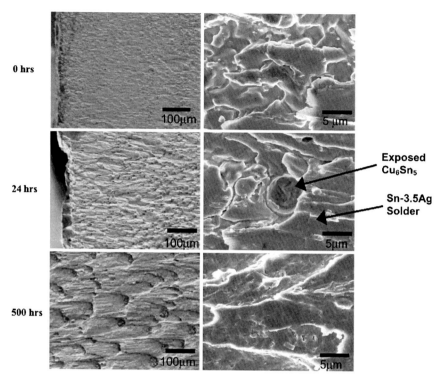

Fig. 1.25 Shear fracture surface of the Sn–3.5Ag/Cu solder joints aged at 175 °C for different times. Reprinted from Ref. [87], with kind permission from Springer Science+Business Media

Table 1.6 Mechanical properties parameters of Cu and IMCs [4, 96, 97]

Properties	Cu_6Sn_5	Cu_3Sn	Cu
Elastic modulus (GPa)	85.56	108.3	110.32
Yield strength (MPa)	–	–	340
Tensile strength (MPa)	–	–	351
Shear modulus (GPa)	50.21	42.41	40.0
Shear strength (MPa)	–	–	–
Fracture toughness (MPa m$^{-1/2}$)	1.4	1.7	–
Hardness (GPa)	3.70	3.36	HV160

According to the tensile and shear fracture behavior of the solder joints, it can be predicted that fracture behavior of the IMC layer is an important factor on adhesive property of the solder joint, while the exact data on the strength of IMC layer is still lacked, as in Table 1.6 [96, 97]. The elastic modulus and hardness of the IMCs can be obtained by nano indentation, but cannot get the strength of the IMCs. Besides, the size of the nano indentation is usually very small, and the loading is quite different from the loadings suffered by the IMC layer in real service condition, thus the nanoindentation test cannot be used to evaluate the reliability of the joint interface. The fracture strength of the IMC layer need to be further investigated through innovative design of the loading method.

1.4.4 Fatigue Damage Behavior of Solder Joint

Since the solder joints usually suffer cyclic loadings, the fatigue resistance of the solder joint is very important. With decrease in size of the circuits, the service condition of the solder joints is more severe and fatigue property of the solder joint is more important. However, because the fatigue test needs a large number of samples and a long period, it is difficult to carry out, and the complex microstructure of the solder joint makes its fatigue damage far more complex than that of the single-phase materials. Therefore, the report lacks on fatigue properties of the solder joints.

The investigations reveal that the fatigue deformation concentrates inside the solder, and the fatigue fracture occurs at the joint interface [98–104]. The macroscopic fatigue fracture surface of the BGA solder joint is shown in Fig. 1.26. As in the figure, obvious deformation occurs inside the solder, while the joint fracture inside the solder close to the joint interface. Since the fatigue damage process is long, the interfacial IMC layer may grow obviously during the fatigue process, and the accumulated strain energy can result in microstructure evolution [103–106]. The interfacial morphologies of the as-soldered and thermal aged joints are shown in Fig. 1.27, in which the IMC thickness increases obviously, and there is coarsen in microstructure of the solder.

At low strain amplitude, sometimes fatigue striations can be observed at some regions of the fatigue fracture surface of the solder joint (see Fig. 1.28), but not as regular as that in the common metallic materials. Besides, as solder is very soft, the

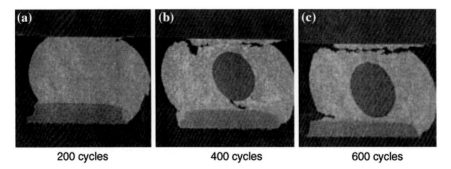

| 200 cycles | 400 cycles | 600 cycles |

Fig. 1.26 Morphologies of solder joint thermal cycled for different cycles. Reprinted from Ref. [98], Copyright 2006, with permission from Elsevier

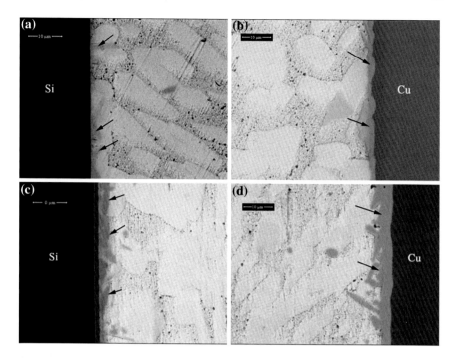

Fig. 1.27 Interfacial morphologies of Si/Sn–52In/Cu solder joint cycled at the temperature range of −40 to 100 °C: **a** on the Si side, as-reflowed, **b** on the Cu side, as-reflowed, **c** on the Si side, after 700 thermal cycles, and **d** on the Cu side, after 700 thermal cycles. Reprinted from Ref. [100], with kind permission from Springer Science+Business Media

fatigue striations formed at early stage may be "wearied down" during the later fatigue process and can also not be found on the final fracture surface.

1.4.5 Creep Behavior of Solder Joints

Creep deformation of the solder joints is contributed mainly by the solder, thus creep resistance of the solder joint depends on the solder. The shear creep stress–strain curve of a series of Pb-free solder/Cu solder joints are shown in Fig. 1.29 [107] in which the creep behavior of the solder joints fit well with Eq. 1.3, and the effect of temperature on creep rate is significant. However, the creep behaviors of the solder and solder joints are still quite different. Similar to the damage under tensile, shear, and fatigue loadings, the creep deformation also tends to concentrate around the joint interface, and final fracture occurs at the joint interface [108, 109], as in Fig. 1.30. Therefore, creep behavior of the solder joints and their resistance to creep damage are also significantly affected by the interfacial microstructure, and need to be evaluated comprehensively based on understandings on creep behavior

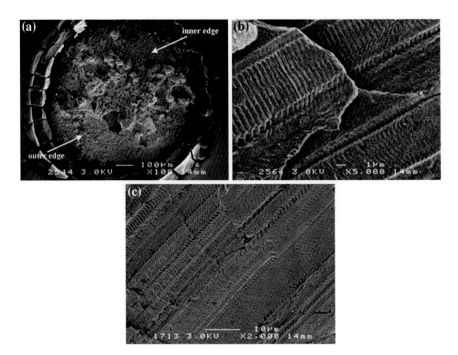

Fig. 1.28 Fracture surfaces on the solder joints thermal cycled at a temperature range of −55 to 150 °C: **a** fracture surface of the Sn–7In–4.1Ag–0.5Cu/Cu solder joint, **b** transgranular failure at the Sn–7In–4.1Ag–0.5Cu/Cu solder joint, and **c** transgranular failure at the 95.5Sn–4Ag–0.5Cu/Cu solder joint. Reprinted from Ref. [97], with kind permission from Springer Science+Business Media

of the solder and fracture behavior of the joint interface. Besides, since the real loading suffered by tsolder joint is creep-fatigue loading, investigations on creep-fatigue behavior of the solder joint is more meaningful for evaluation of the reliability.

1.4.6   Problems in Research

The mechanical loadings suffered by the solder joints is a combined loading of tensile, shear, creep, and fatigue, and their damage behavior is affected by microstructure, temperature, and strain rate, thus the damage behavior is extremely complex. Thus far the investigations on the Pb-free solders and solder joints have got amount of data, but the test conditions of different references are quite different and difficult to make a comparison. Compared with the perfect data and evaluation system of the Sn–37Pb solder, investigation on mechanical properties of the Pb-free solder is still lacking.

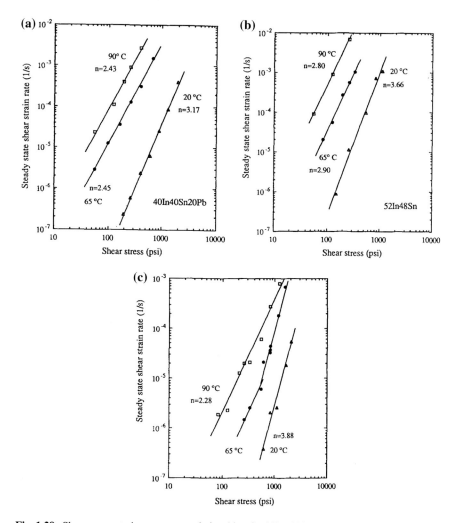

Fig. 1.29 Shear creep strain rate–stress relationship of **a** 40In–40Sn–20Pb, **b** 52In–48Sn, **c** 43Sn–43Pb–14Bi solder joints at 20, 65 and 90 °C. Reprinted from Ref. [107], with kind permission from Springer Science+Business Media

1.5   Research Content and Purpose of This Dissertation

Based on the discussions on investigations of the Pb-free soldering and the mechanical properties of the solder joints, it can be found that related investigations mainly focus on the following three directions:

(1) Mechanical property of the Pb-free solder: The mechanical property is an important factor for selection of the solder, the comprehensive research on mechanical property of the solder has been carried out during the development

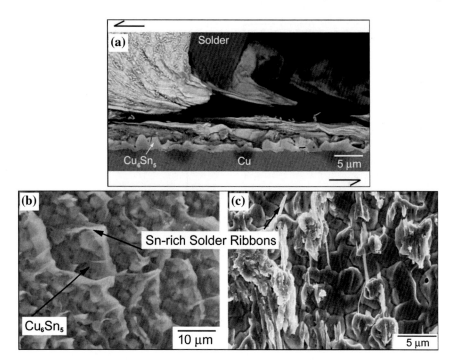

Fig. 1.30 Creep behaviors of Sn–3.5Ag/Cu solder joints: **a** strain localization and fracture around the Cu_6Sn_5/solder interface; creep fracture morphology at (**b**) 95 °C, decohesion of the solder from the Cu_6Sn_5 intermetallic (**c**) and 130 °C, oxide formation is observed in pure Sn. Reprinted from Ref. [108], Copyright 2004, with permission from Elsevier

of solder, and abundant data are obtained. However, the mechanical property of the solder is only one influencing factor on reliability of the solder joints, and the test loadings on solders in most studies are different from the real loadings suffered by the solder joint. Therefore, this study will pay attention to the deformation of the solder in the solder joint rather than the solder itself, to make the test result more useful for evaluation of the reliability of the solder joints.

(2) Interfacial reaction and IMC growth behavior at the solder joint: As the IMC layer at the joint interface shows significant influence on mechanical properties of the solder joints, the solder joint can be optimized through controlling the growth of the interfacial IMC layer. The effective method to restrain the growth of IMC is adding some alloy elements in the solder and optimizing the soldering parameters. Investigations on the growth behavior of interfacial IMCs are already quite intensive, while the investigations on their fracture behavior and mechanical properties are lacking. In this study, not only the growth behavior of the IMCs will be studied, the fracture behaviors of the IMC layers will also be investigated comprehensively.

(3) Mechanical properties of the solder joint interface: The investigation on this field is insufficient. According to weak points of studies on mechanical property of the Pb-free soldering, in this study some innovative designs on mechanical property test method of the solder joints are applied, and the tensile-compress fatigue, creep-fatigue, thermal fatigue behaviors of the solder joints are studied, the damage behaviors of the solder joints at different conditions and the influences of the interfacial microstructure, strain, stress, and temperature are discussed.

Specifically, in this study the following topics are studied:

- Fracture behavior of interfacial IMC layers at Cu/Pb-free interface.
- Tensile-compress fatigue damage behavior of Cu/Pb-free solder joints.
- Creep-fatigue behavior of Cu/Pb-free solder joints.
- Thermal fatigue behavior of Cu/Pb-free solder joints.

References

1. Greig WJ. Integrated circuit packaging, assembly and interconnections (Springer series in advanced microelectronics). New York: Springer; 2007.
2. Laurila T, Vuorinen V, Kivilahti JK. Interfacial reactions between lead-free solders and common base materials. Mater Sci Eng R. 2005;49:1–60.
3. Tu KN. Solder joint technology: materials, properties, and reliability. New York: Springer; 2007. p. 386.
4. Abtew M, Selvaduray G. Lead-free solders in microelectronics. Mater Sci Eng R. 2000;27:95–141.
5. Zeng K, Tu KN. Six cases of reliability study of Pb-free solder joints in electronic packaging technology. Mater Sci Eng R. 2002;38:55–105.
6. Massalski TB, Okamoto H. Binary alloy phase diagrams. 2nd ed. New York: ASM International; 1990.
7. McCormack M, Jin S, Kammlott GW, Chen HS. New Pb-free solder alloy with superior mechanical properties. Appl Phys Lett. 1993;63:15–7.
8. Ventura T, Terzi S, Rappaz M, Dahle AK. Effects of solidification kinetics on microstructure formation in binary Sn–Cu solder alloys. Acta Mater. 2011;59:1651–8.
9. Grossmann G, Tharian J, Jud P, Sennhauser U. Microstructural investigation of lead-free BGAs soldered with tin-lead solder. Solder Surf Mt Technol. 2005;17:10–21.
10. Loomans ME, Fine ME. Tin-silver-copper eutectic temperature and composition. Metall Mater Trans A. 2000;31:1155–62.
11. Moon KW, Boettinger WJ, Kattner UR, Biancaniello FS, Handwerker CA. Experimental and thermodynamic assessment of Sn–Ag–Cu solder alloys. J Electron Mater. 2000;29:1122–36.
12. Mei Z, Morris JW Jr. Characterization of eutectic Sn–Bi solder joints. J Electron Mater. 1992;21:599–607.
13. Glazer J. Microstructure and mechanical properties of Pb-free solder alloys for low-cost electronic assembly: a review. J Electron Mater. 1994;23:693–700.
14. Liu PL, Shang JK. Interfacial embrittlement by bismuth segregation in copper/tin–bismuth Pb-free solder interconnect. J Mater Res. 2001;16:1651–9.

15. Lin HJ, Chuang TH. Intermetallic reactions in reflowed and aged Sn–9Zn solder ball grid array packages with Au/Ni/Cu and Ag/Cu pads. J Electron Mater. 2006;35:154–64.
16. Laurila T, Vuorinen V, Paulasto-Kröckel M. Impurity and alloying effects on interfacial reaction layers in Pb-free soldering. Mater Sci Eng R. 2010;68:1–38.
17. Kang JS, Gagliano RA, Ghosh G, Fine ME. Isothermal solidification of Cu/Sn diffusion couples to form thin-solder joints. J Electron Mater. 2002;31:1238–43.
18. Larsson AK, Stenberg L, Lidin S. The superstructure of domain-twinned eta'-Cu6Sn5. Acta Crystallogr B. 1994;50:636–43.
19. Larsson AK, Stenberg L, Lidin S. Crystal-structure modulations in eta-Cu5Sn4. Z Kristallogr. 1995;210:832–7.
20. Gao F, Nishikawa H, Takemoto T. Intermetallics evolution in Sn–3.5Ag based lead-free solder matrix on an OSPCu finish. J Electron Mater. 2007;36:1630–4.
21. Yoon JW, Lim JH, Lee HJ, Joo J, Jung SB, Moon WC. Interfacial reactions and joint strength of Sn–37Pb and Sn–3.5Ag solders with immersion Ag-plated Cu substrate during aging at 150 °C. J Mater Res. 2006;21:3196–204.
22. Tseng HW, Liu CY. Evolution of Ag3Sn compound formation in Ni/Sn5Ag/Cu solder joint. Mater Lett. 2008;62:3887–9.
23. Song JM, Lin JJ, Huang CF, Chuang HY. Crystallization, morphology and distribution of Ag₃Sn in Sn–Ag–Cu alloys and their influence on the vibration fracture properties. Mater Sci Eng A. 2007;466:9–17.
24. Henderson DW, Gosselin T, Sarkhel A, Kang SK, Choi WK, Shih DY, Goldsmith C, Puttlitz KJ. Ag₃Sn plate formation in the solidification of near ternary eutectic Sn–Ag–Cu alloys. J Mater Res. 2002;17:2775–8.
25. Zou HF, Yang HJ, Tan J, Zhang ZF. Preferential growth and orientation relationship of Ag₃Sn grains formed between molten Sn and (001) Ag single crystal. J Mater Res. 2009;24:2141–4.
26. Gur D, Bamberger M. Reactive isothermal solidification in the Ni–Sn system. Acta Mater. 1998;46:4917–23.
27. Ghosh G. Interfacial microstructure and the kinetics of interfacial reaction in diffusion couples between Sn–Pb solder and Cu/Ni/Pd metallization. Acta Mater. 2000;48:3719–38.
28. Görlich J, Baither D, Schmitz G. Reaction kinetics of Ni/Sn soldering reaction. Acta Mater. 2010;58:3187–97.
29. Li JF, Mannan SH, Clode MP, Chen K, Whalley DC, Liu C, Hutt DA. Comparison of interfacial reactions of Ni and Ni–P in extended contact with liquid Sn–Bi-based solders. Acta Mater. 2007;55:737–52.
30. Bader S, Gust W, Hieber H. Rapid formation of intermetallic compounds by interdiffusion in the Cu–Sn and Ni–Sn systems. Acta Metall Mater. 1995;43:329–37.
31. Dybkov VI. Effect of dissolution on the Ni3Sn4 growth kinetics at the interface of Ni and liquid Sn-base solders. Solid State Phenom. 2008;138:153–8.
32. Chan YC, Yang D. Failure mechanisms of solder interconnects under current stressing in advanced electronic packages. Prog Mater Sci. 2010;55:428–75.
33. Evans JW. A guide to lead-free solders. 1st ed. London: Springer; 2005.
34. Ohguchi KI, Sasaki K, Ishibashi M. A quantitative evaluation of time-independent and time-dependent deformations of lead-free and lead-containing solder alloys. J Electron Mater. 2006;35:132–9.
35. Park S, Dhakal R, Lehman L, Cotts E. Measurement of deformations in SnAgCu solder interconnects under in situ thermal loading. Acta Mater. 2007;55:3253–60.
36. Frear DR. The mechanical behavior of interconnect materials for electronic packaging. J Mater. 1996;48:49–53.
37. Glazer J. Metallurgy of low temperature Pb-free solders for electronic assembly. Int Mater Rev. 1995;40:65–93.
38. Thwaites CJ. Soft soldering handbook. International Tin Research Institute, Publication No. 533; 1977.

39. Yamagishi Y, Ochiai M, Ueda H, Nakanishi T, Kitazima M. Pb-free solder of Sn–58Bi improved with Ag. In: Proceedings of the 9th international microelectronics conference, Omiya, Japan. 1996. pp. 252–5.
40. Tojima K. Wetting characteristics of lead-free solders, senior project report. Materials Engineering Department, San Jose State University, May 1999.
41. Hua F, Glazer J. Lead-free solders for electronic assembly, design and reliability of solders and solder interconnections. In: Mahidhara RK, Frear DR, Sastry SML, Liaw KL, Winterbottom WL, editors. The minerals, metals and materials society. 1997. pp. 65–74.
42. Andersson C, Sun P, Liu JH. Tensile properties and microstructural characterization of Sn–0.7Cu–0.4Co bulk solder alloy for electronics applications. J Alloys Compd. 2008;457:97–105.
43. Shohji I, Yoshida T, Takahashi T, Hioki S. Tensile properties of Sn–Ag based lead-free solders and strain rate sensitivity. Mater Sci Eng A. 2004;366:50–5.
44. Fouassier O, Heintz JM, Chazelas J, Geffroy PM, Silvain JF. Microstructural evolution and mechanical properties of SnAgCu alloys. J Appl Phys. 2006;100:043519.
45. Zhu FL, Zhang HH, Guan RF, Liu S. Effects of temperature and strain rate on mechanical property of Sn96.5Ag3Cu0.5. Microelectron Eng. 2007;84:144–50.
46. Mavoori H, Chin J, Vayman S, Moran B, Keer L, Fine M. Creep, stress relaxation, and plastic deformation in Sn–Ag and Sn–Zn eutectic solders. J Electron Mater. 1997;26:783–90.
47. Ochoa F, Willlams JJ, Chawla N. Effects of cooling rate on the microstructure and tensile behavior of a Sn–3.5wt.%Ag solder. J Electron Mater. 2003;32:1414–20.
48. Schoeller H, Bansal S, Knobloch A, Shaddock D, Cho J. Microstructure evolution and the constitutive relations of high-temperature solders. J Electron Mater. 2009;38:802–9.
49. ASM International. Electronic materials handbook, vol. 1. Packaging Materials Park, OH: ASM International; 1989. p. 640.
50. Solder alloy data: mechanical properties of solders and soldered joints. International Tin Research Institute, Uxbridge, England, p. 60.
51. Tomlinson WJ, Collier I. The mechanical properties and microstructures of copper and brass joints soldered with eutectic tin-bismuth solder. J Mater Sci. 1987;22:1835–9.
52. Artaki I, Jackson AM, Vianco PT. Evaluation of lead-free joints in electronic assemblies. J Electron Mater. 1994;23:757–64.
53. Ma HT. Constitutive models of creep for lead-free solders. J Mater Sci. 2009;44:3841–51.
54. Morris JW Jr, Goldstein JLF, Mei Z. Microstructure and mechanical properties of Sn–In and Sn–Bi solders. J Electron Mater. 1993;22:25–7.
55. Mukherjee AK, Bird JE, Dorn JE. Experimental correlations for high-temperature creep. Trans Am Soc Met. 1969;62:155–79.
56. Hertzberg RW. Deformation and fracture mechanics of engineering materials. 4th ed. New York: Wiley; 1996.
57. Chen ZG, Shi YW. Xia ZD. Constitutive relations on creep for SnAgCuRE lead-free solder joints. 2004;33:964–71.
58. Zhang KK, Lwang Y, Fan YL, Zhang X. Research on creep properties of Sn2.5Ag0.7CuXRE lead-free soldered joints for surface mount technology. Mater Sci. 2007;353–358:2912–5.
59. Ma HT, Suhling JC. A review of mechanical properties of lead-free solders for electronic packaging. J Mater Sci. 2009;44:1141–58.
60. Yan YF, Ji LQ, Zhang KK, Yan HX, Feng LF. Foundation of steady state creep constitutive equation of SnCu soldered joints. Trans China Weld Inst. 2007;28(9):75–9.
61. Igoshev VI, Kleiman JI. Creep phenomena in lead-free solders. J Electron Mater. 2000;29:244–50.
62. Xiao Q, Armstrong WD. Tensile creep and microstructural characterization of bulk Sn3.9Ag0.6Cu lead-free solder. J Electron Mater. 2005;34:196–211.
63. Wiese S, Schubert A, Walter H, Dudek R, Feustel F, Meusel E, Michel B. Constitutive behavior of lead-free solders vs. lead containing solders–experiments on bulk specimens and

flip-chip joints. In: Proceeding of the 51st electronic components and technology conference, pp. 890–902.

64. Clech JP. Review and analysis of lead-free materials properties, NIST. Available at http://www.metallurgy.nist.gov/solder/clech/Sn-Ag-Cu_Main.htm.

65. Vianco PT. Fatigue and creep of lead-free solder alloys: fundamental properties. 1st edn. New York: ASM International; 2006.

66. Pang JHL, Xiong BS, Low TH. Creep and fatigue characterization of lead-free 95.5Sn–3.8Ag–0.7Cu solder. In: Proceeding of 54th electronic components and technology conference. 2004. pp. 1333–7.

67. Shi YW, Liu JP, Yan YF, Xia ZD, Lei YP, Guo F, Li XY. Creep properties of composite solders reinforced with nano- and microsized particles. J Electron Mater. 2008;37:507–17.

68. Shi YW, Liu JP, Xia ZD, Lei YP, Guo F, Li XY. Creep property of composite solders reinforced by nano-sized particles. J Mater Sci: Mater Electron. 2008;19:349–56.

69. Guo F, Lee J, Lucas JP, Subramanian KN, Bieler TR. Creep properties of eutectic Sn–3.5Ag solder joints reinforced with mechanically incorporated Ni particles. J Electron Mater. 2001;30:1222–7.

70. Tai F, Guo F, Liu JP, Xia ZD, Shi YW, Lei YP, Li XY. Creep properties of Sn–0.7Cu composite solder joints reinforced with nano-sized Ag particles. Solder Surf Mt Technol. 2010;22:50–6.

71. Nozaki M, Sakane M, Tsukada Y. Crack propagation behavior of Sn–3.5Ag solder in low cycle fatigue. Int J Fatigue. 2008;30:1729–36.

72. Arfaei B, Cotts E. Correlations between the microstructure and fatigue life of near-eutectic Sn–Ag–Cu Pb-free solders. J Electron Mater. 2009;38:2617–27.

73. Zhao J, Mutoh Y, Miyashita Y, Wang L. Fatigue crack growth behavior of Sn–Pb and Sn-based lead-free solders. Eng Fract Mech. 2003;70:2187–97.

74. Kanchanomai C, Miyashita Y, Mutoh Y, Mannan SL. Influence of frequency on low cycle fatigue behavior of Pb-free solder 96.5Sn–3.5Ag. Mater Sci Eng A. 2003;345:90–8.

75. Kanchanomai C, Mutoh Y. Low-cycle fatigue prediction model for Pb-free solder 96.5Sn–3.5Ag. J Electron Mater. 2004;33:329–33.

76. Pang JHL, Xiong BS, Low TH. Low cycle fatigue study of lead free 99.3Sn–0.7Cu solder alloy. Int J Fatigue. 2004;26:865–72.

77. Pang JHL, Xiong BS, Low TH. Low cycle fatigue models for lead-free solders. Thin Solid Films. 2004;462–463:408–12.

78. Shang JK, Zeng QL, Zhang L, Zhu QS. Mechanical fatigue of Sn-rich Pb-free solder alloys. J Mater Sci: Mater Electron. 2007;18:211–27.

79. Takemoto T, Matsunawa A, Takahashi M. Tensile test for estimation of thermal fatigue properties of solder alloys. J Mater Sci. 1997;32:4077–84.

80. Lea C. A scientific guide to surface mount technology. GB-Port Erin, British Isles: Electrochemical Publications Ltd; 1988.

81. Kikuchi S, Nishimura M, Suetsugu K, Ikari T, Matsushige K. Strength of bonding interface in lead-free Sn alloy solders. Mater Sci Eng A. 2001;319–321:475–9.

82. Lee HT, Chen MH, Jao HM, Liao TL. Influence of interfacial intermetallic compound on fracture behavior of solder joints. Mater Sci Eng A. 2003;358:134–41.

83. Lee HT, Lee YH. Adhesive strength and tensile fracture of Ni particle enhanced Sn–Ag composite solder joints. Mater Sci Eng A. 2006;419:172–80.

84. Zou HF, Zhu QS, Zhang ZF. Growth kinetics of intermetallic compounds and tensile properties of Sn–Ag–Cu/Ag single crystal joint. J Alloy Compd. 2008;461:410–7.

85. Dao M, Chollacoop N, Van Vliet KJ, Venkatesh TA, Suresh S. Computational modeling of the forward and reverse problems in instrumented sharp indentation. Acta Mater. 2001;49:3899–918.

86. Deng X, Chawla N, Chawla KK, Koopman M. Deformation behavior of (Cu, Ag)–Sn intermetallics by nanoindentation. Acta Mater. 2004;52:4291–303.

87. Deng X, Sidhu RS, Johnson P, Chawla N. Influence of reflow and thermal aging on the shear strength and fracture behavior of Sn–3.5Ag solder/Cu joints. Metall Mater Trans A. 2005;36A:55–64.
88. Kima KS, Huh SH, Suganuma K. Effects of intermetallic compounds on properties of Sn–Ag–Cu lead-free soldered joints. J Alloys Compd. 2003;352:226–36.
89. Yoon JW, Kim SW, Jung SB. Interfacial reaction and mechanical properties of eutectic Sn–0.7Cu/Ni BGA solder joints during isothermal long-term aging. J Alloys Compd. 2005;391:82–9.
90. Choi WK, Kim JH, Jeong SW, Lee HM. Interfacial microstructure and joint strength of Sn–3.5Ag-X (X = Cu, In, Ni) solder joint. J Mater Res. 2002;17:43–51.
91. Kim SW, Yoon JW, Jung SB. Interfacial reactions and shear strengths between Sn–Ag-based Pb-free solder balls and Au/EN/Cu metallization. J Electron Mater. 2004;33:1182–9.
92. Ahat S, Sheng M, Le L. Microstructure and shear strength evolution of SnAg/Cu surface mount solder joint during aging. J Electron Mater. 2001;30:1317–22.
93. Lee YH, Lee HT. Shear strength and interfacial microstructure of Sn–Ag–xNi/Cu single shear lap solder joints. Mater Sci Eng A. 2007;444:75–83.
94. Anderson IE, Harringa JL. Elevated temperature aging of solder joints based on Sn–Ag–Cu: effects on joint microstructure and shear strength. J Electron Mater. 2004;33:1485–96.
95. Zou HF, Zhang ZF. Ductile-to-brittle transition induced by increasing strain rate in Sn–3Cu/Cu joints. J Mater Res. 2008;23:1614–7.
96. Fields RJ, Low SR. Physical and mechanical properties of intermetallic compounds commonly found in solder joints. Metallurgy Division of National Institute of Standards and Technology (NIST), USA, Technical paper, 2001.
97. Frear DR, Burchett SN, Morgan HS, Lau JH. The mechanics of solder alloy interconnects. 1st ed. New York: Springer; 1994.
98. Wang ZX, Dutta I, Majumdara BS. Thermal cycle response of a lead-free solder reinforced with adaptive shape memory alloy. Mater Sci Eng A. 2006;421:133–42.
99. Towashiraporn P, Gall K, Subbarayan G, McIlvanie B, Hunter BC, Love D, Sullivan B. Power cycling thermal fatigue of Sn–Pb solder joints on a chip scale package. Inter J Fatigue. 2004;26:497–510.
100. Guo HY, Guo JD, Shang JK. Influence of thermal cycling on the thermal resistance of solder interfaces. J Electron Mater. 2009;38:2470–8.
101. Kobayashi T, Lee A, Subramanian KN. Impact behavior of thermomechanically fatigued Sn-based solder joints. J Electron Mater. 2009;38:2659–67.
102. Seah SKW, Wonga EH, Shim VPW. Fatigue crack propagation behavior of lead-free solder joints under high-strain-rate cyclic loading. Script Mater. 2008;59:1239–42.
103. Lehman LP, Xing Y, Bieler TR, Cotts EJ. Cyclic twin nucleation in tin-based solder alloys. Acta Mater. 2010;58:3546–56.
104. Lee KO, Yu J, Park TS, Lee SB. Low-cycle fatigue characteristics of Sn-based solder joints. J Electron Mater. 2004;33:249–57.
105. Sundelin JJ, Nurmib ST, Lepistö TK. Recrystallization behaviour of SnAgCu solder joints. Mater Sci Eng A. 2008;474:201–7.
106. Erinc M, Assman TM, Schreurs PJG, Geers MGD. Fatigue fracture of SnAgCu solder joints by microstructural modeling. Int J Fract. 2008;152:37–49.
107. Mei Z, Morris JWJR. Superplastic creep of low melting point solder joints. J Electron Mater. 1992;21:401–7.
108. Kerr M, Chawla N. Creep deformation behavior of Sn–3.5Ag solder/Cu couple at small length scales. Acta Mater. 2004;52:4527–35.
109. Telang AU, Bieler TR. The orientation imaging microscopy of lead-free Sn–Ag solder joints. JOM. 2005: 44–9.

Chapter 2
Fracture Behavior of IMCs at Cu/Pb-Free Solder Interface

2.1 Introduction

During the soldering process, the molten solder usually forms IMCs with the substrate material, which provides thermal, electrical, and mechanical integrity for the solder joint [1, 2]; therefore, it is necessary for the formation of high-quality solder joints. However, the IMCs are usually very brittle [3] and their thicknesses increase gradually when the solder joints generate heat during the service period; brittle fracture can easily occur in the IMCs and their strength can also get decreased [4–6]. So far, there have been many investigations trying to improve the reliability of the solder joints through decreasing the growth rate of the IMCs, but they can hardly eliminate the brittle fracture inside the IMC layer. With the decrease in size of the solder joints, the influence of IMC layer on the adhesive property of the solder joint is more significant, and the understanding of the fracture behavior of IMCs is more meaningful for evaluation of the reliability of the solder joints.

Because the IMC layers in the solder joints are usually very thin, it is difficult to carry out traditional mechanical property tests on the IMCs; the major method to test their mechanical property is the nanoindentation test, which can get the elastic modulus and hardness of the IMC [5–10], but cannot get their strength or evaluate their fracture behavior. Jiang et al. [11] reported a micropillar compression test to assess the mechanical properties of the IMCs, which is important because it is the first report describing the strength and fracture mechanisms of Cu_6Sn_5; while fracture of the Cu_6Sn_5 under compression loading is still a little bit different from the fracture in the service condition, the IMC grains in the real solder joints are usually elongated hemispherical or "scallop shaped", and suffer the shear loading.

So far investigations on the fracture behavior of the IMC layers under external loadings are abundant, and some qualitative conclusions have been obtained. It was found that with increases in aging time and thickness of the IMC layer, the fracture transforms from a ductile fracture inside the solder into a cleavage fracture inside the IMC layer, and the joint strength decreases obviously [12–15]. However,

© Springer-Verlag Berlin Heidelberg 2016
Q. Zhang, *Investigations on Microstructure and Mechanical Properties of the Cu/Pb-free Solder Joint Interfaces*, Springer Theses, DOI 10.1007/978-3-662-48823-2_2

generally these investigations focus on the evolution of joint strength with increasing aging time rather than the fracture behavior of the IMCs. Besides, very few studies have investigated the influence of reflow time on the fracture behavior of the IMC layer and the solder joint. In fact, understandings on the influence of reflow time are more meaningful for finding proper soldering parameters, because the reflow time can be controlled to get the desired interfacial IMC thickness.

As the interfacial IMC layers are very thin ($\sim$ 1–10 μm) and located around the solder–substrate interface, when the solder joint is deformed by external loading, not only the loadings suffered by the solder can result in the fracture of IMC, but also the deformation of the substrate may also cause the fracture [16, 17]. Since the solder joints in the microelectronic devices serve in increasingly complex and dynamic environments, possibility of the fracture induced by deformation of the substrate is increasing, while the investigations on such cases are lack.

For the aforementioned reasons, in this chapter some new methods were designed to investigate the fracture behavior of the IMC layers induced by shear stress and deformation of the solder and the substrate, and the influencing factors on fracture behavior of the IMC layer are discussed. The research progress will be helpful for evaluating the reliability of the solder joints, and it is hoped that the proposed new test methods better reveal the property of the IMC layers.

2.2 Experimental Procedure

2.2.1 Indentation Test of Interfacial IMC Grains

The substrate material used for the indentation test of the interfacial IMC grains is cold-drawn polycrystalline Cu with a purity of 99.99 % and a yield strength of about 300 MPa. The solder used in this study is Sn–4Ag solder alloy prepared by melting high-purity (>99.99 %) tin and silver at 800 °C for 30 min in vacuum. To prepare the test samples, a small block (5 mm × 5 mm × 5 mm) was first spark cut from the Cu substrate, and then its side surface for reflowing was ground and carefully polished. After air drying, a soldering flux was dispersed on the polished area and a piece of solder alloy was stuck on it. The prepared samples were put in an oven with a temperature of 260 °C for 8 min after melting of the solder and then air cooled.

After the reflowing process, a thin sheet with a thickness of about 1.5 mm was sliced from the sample and its side surface was ground and carefully polished. To expose the Cu_6Sn_5 grains, the superficial Sn–4Ag solder around the joint interface was removed by corrosion using the 5 % HCl + 3 % HNO_3 + CH_3OH (wt%) etching solution and the morphologies of the target Cu_6Sn_5 grains were observed by a ZEISS Supra 35 scanning electron microscope (SEM), in order to make a morphological comparison of the indentation test. The morphology of the joint interface after corrosion is shown in Fig. 2.1a.

Fig. 2.1 **a** Morphologies of
the Cu_6Sn_5 grains at the
Sn–4Ag/Cu interface,
b illustration on the shear test
by Dynamic Ultramicroscopic
Hardness tester. Reprinted
from Ref. [18]. Copyright
2011, with permission from
AIP Publishing LLC

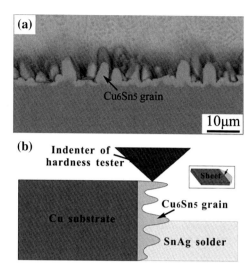

The DUH-211 Dynamic Ultramicroscopic Hardness tester was employed to conduct the indentation tests, because it is easily operated and can record the dynamic load–depth relationships. The indenter is a triangular pyramid and the indentation location is the center portion of the target Cu_6Sn_5 grains, as illustrated in Fig. 2.1b. The load was chosen to be 20 mN and the loading speed was set to be 1.90 mN/s. The force–depth relationships were recorded. After the indentation tests, the front views of the target Cu_6Sn_5 grains were first observed, the cracked Cu_6Sn_5 fragments were flushed away by ultrasonic cleaning, and then the fracture surface of the target grains was also observed by SEM.

2.2.2 Fracture of IMCs Induced by Deformation of Solder

The substrate material used for the in situ observation of the fracture behavior of the interfacial IMCs induced by deformation of the solder is also cold-drawn polycrystalline Cu, and the solder is the Sn–4Ag (wt%) alloy. The sample preparation process for interfacial observation is similar to the process mentioned in Sect. 2.2.1, as in Fig. 2.2a. The prepared samples were put in an oven with a temperature of 260 °C, for 1, 3, or 8 min; the molten solder is then taken out and quenched with alcohol. After the reflow process, the overlaid solders on the samples were eroded with the 5 % HCl + 3 % HNO_3 + CH_3OH (wt%) etching solution to expose the Cu_6Sn_5 grains. The morphologies of the Cu_6Sn_5 layers were observed by a ZEISS Supra 35 SEM and a LEXT OLS4000 measuring laser confocal microscope and their roughnesses were measured directly using the latter. The contours of the interfacial IMC layers were analyzed using the SISC IAS V8.0 software to obtain their thickness.

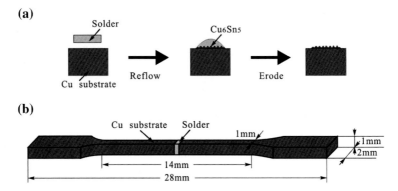

Fig. 2.2 **a** Preparing process of samples for interfacial IMC morphology observation, **b** shape and dimension of the tensile samples. Reprinted from Ref. [19]. Copyright 2012, with permission from AIP Publishing LLC

To prepare the tensile samples, the Cu substrates were first spark cut into blocks with a dog-bone-shaped side cross section; the blocks were then spark cut at their midst and the surfaces for reflow were ground and electrolytically polished. A flux was dispersed on the polished surfaces of the blocks before reflow, then the Sn–4Ag sheets were sandwiched between them, and finally two graphite plates were clamped on the sides to avoid outflow of the molten solder. The reflow processes are the same to those mentioned above. Then the prepared samples were sliced into standard tensile specimens, their side surfaces were ground with 2000# SiC abrasive paper, and carefully polished with 1-μm diamond powder. The shape and dimension of the final prepared specimens are presented in Fig. 2.2b.

The tensile tests were carried out by Gatan MTEST2000ES tensile stage installed on the ZEISS Supra 35 SEM. The motor speeds were set as 0.033, 0.1, and 0.4 mm min^{-1}, respectively. Since the Cu substrate only shows very slight elastic deformation, the strain rate was approximately calculated by dividing the cross-beam speed with the thickness of the solder in the solder joints ($\sim$0.5 mm), thus the strain rates are estimated to be about 1.1×10^{-3} s^{-1}, 3.3×10^{-3} s^{-1} and 1.33×10^{-2} s^{-1}, respectively. Three samples were tested under each condition and the tensile strength is their average value. The interfacial fracture morphologies of selected solder joints were observed at certain strains, and the fracture surfaces were also observed by SEM after the tests to comprehensively reveal the fracture mechanisms.

2.2.3 Fracture of IMCs Induced by Deformation of Substrate

Two kinds of Cu substrates were used in the study of fracture behavior of IMCs induced by deformation of the substrate. The first is the cold-drawn oxygen-free-high-conductivity (OFHC) Cu with a yield strength of about 300 MPa. The second one is Cu crystal composed of centimeter-sized grains with a yield strength of about 32 MPa. Since the coarse grains are close to the prepared specimens in size, the latter can be approximately regarded as single crystals. The solder used in this study is also the Sn–4Ag (wt%) alloy.

To prepare the test specimens, the Cu substrates were first spark cut into blocks with a side section of dog-bone shape, and then the surfaces for soldering were ground and electrolytically polished. After air drying, a soldering flux was dispersed on the polished area, and a thin solder sheet was stick on it. The prepared samples were put in an oven with a temperature of 260 °C, kept for 8 min after melting of the solder, and then cooled down in air. As the fracture behaviors of the IMC layer are affected by its thickness, some samples were aged at 180 °C for 4 days to get thicker IMC layer. After the reflow and aging processes, the prepared samples were sliced into test specimens and then their side surfaces were ground and carefully polished for interfacial observations. The preparing process and dimension of the test specimens are presented in Fig. 2.3. For the specimens in this study, the IMC layers are driven to deform by the Cu substrate and the solder is unconstrained, in order to make sure that the solder has little influence on fracture of the IMC layers.

All the tests were carried out by the Gatan MTEST2000ES Tensile Stage equipped on the ZEISS Supra 35 SEM, and the crossbeam speed was set as 0.033 mm min^{-1}. To reveal the fracture behaviors of the IMC layers, the tests were paused at some displacements and the interfacial morphologies were observed. As there is no strain gauge on the tensile stage, the macroscopic images comprising the observation region were taken for calculating the strain. The local displacements (Δl) in the macroscopic images were obtained through measuring the distances between the reference points and comparing them with the original distance (l_0), and the local strains were calculated by dividing the displacement with the original distance. The total displacement (ΔL) was recorded by the tensile stage, and the

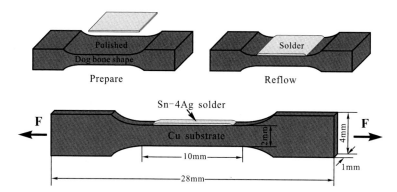

Fig. 2.3 Preparing processes and dimension of the test specimens for observation on fracture behavior of interfacial IMC layers induced by deformation of the substrate. Reprinted from Ref. [20]. Copyright 2011, with permission from Elsevier

equivalent "gauge length" (L_0) was estimated by the equation: $L_0 = l_0 \Delta L / \Delta l$. Based on that, the strain in the stress–strain curves was calculated by dividing ΔL with the mean value of L_0 obtained at different displacements. After the tests, the overlaid solders on the specimens were removed by eroding with a solution composed of 5 % hydrochloric acid, 3 % nitric acid, and 92 % methanol, and the morphologies of the fractured IMC grains were also observed by SEM.

2.3 Shear Fracture Behavior and Strength of Cu_6Sn_5 Grains

A group of force–depth curves of the indentation tests are shown in Fig. 2.4; the depth in the figure is actually the shear displacement of the Cu_6Sn_5 grains. As in the figure, though the loadings are 20 mN for all the tests, the displacements are quite different because the Cu_6Sn_5 grains are different in size and shape. Nevertheless, there are similarities in all the curves. Within each curve, the depth increases approximately linear with increasing force during the initial deforming process. When the force increases to a certain value, a depth burst occurs which may correspond to the fracture in the Cu_6Sn_5 grain [13]. Besides, similar bursts have been observed when the Cu_6Sn_5 micropillar fracture under compression loadings by cleavage [11]. After the burst, the depth again increases with increasing force and the slope is similar to the initial stage. For some curves, there are secondary bursts. As the bursts on different curves are quite different, the fracture morphologies of the correlating Cu_6Sn_5 grains were observed to reveal the differences.

Figures 2.5 and 2.6 show the morphologies of two representative Cu_6Sn_5 grains before and after the indentation tests. The morphologies of a Cu_6Sn_5 grain fractured at a low force are shown in Fig. 2.5. Figure 2.5a exhibits the target Cu_6Sn_5 grain before the indentation test, which is bamboo shoot-like in shape and the length is

Fig. 2.4 Force–depth curves of the shear tests. Reprinted from Ref. [18]. Copyright 2011, with permission from AIP Publishing LLC

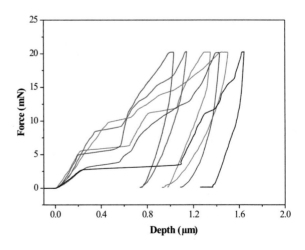

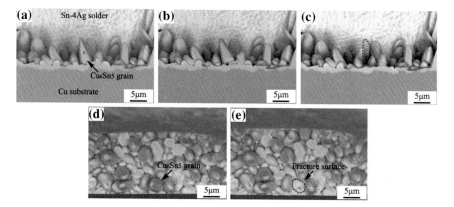

Fig. 2.5 Morphologies of a Cu$_6$Sn$_5$ grain fractured at the foundation: *front views* **a** before indentation test, **b** after indentation test, and **c** after ultrasonic cleaning; *top views* **d** before indentation test and **e** after ultrasonic cleaning. Reprinted from Ref. [18]. Copyright 2011, with permission from AIP Publishing LLC

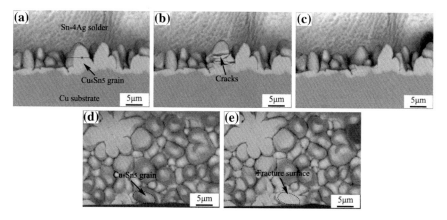

Fig. 2.6 Morphologies of a Cu$_6$Sn$_5$ grain fractured at the center portion: *front views* **a** before indentation test, **b** after indentation test, and **c** after ultrasonic cleaning; *top views* **d** before indentation test and **e** after ultrasonic cleaning. Reprinted from Ref. [18]. Copyright 2011, with permission from AIP Publishing LLC

about 10 μm; the indentation location is indicated by a red "X". After the test, the target Cu$_6$Sn$_5$ grain only exhibited a little tilt compared with the initial state, while the surrounding little Cu$_6$Sn$_5$ grains were broken (see Fig. 2.5b). Figure 2.5c shows the front view after the ultrasonic cleaning. The target Cu$_6$Sn$_5$ was washed away, indicating that actually it has fractured at the foundation under the indentation test. The top views of the target Cu$_6$Sn$_5$ grain are presented in Fig. 2.5d, e, respectively. By comparing them, the fracture surface of the target Cu$_6$Sn$_5$ grain can be determined easily, as indicated by the red circle.

Figure 2.6 shows the morphologies of a Cu_6Sn_5 grain fractured at a higher force. The front view of the target grain before the test is shown in Fig. 2.6a. As in the figure, the target grain is a little bit podgy compared with that in Fig. 2.5a, and the indentation location is also indicated by the red "X". After the indentation test, it is interesting to find that there are some parallel cracks in the target Cu_6Sn_5 grain, but there is no breakage in the surrounding Cu_6Sn_5 grains (see Fig. 2.6b). According to the width and location of the cracks, the cracking at the center portion of the grain is predicated to be formed initially, i.e. it is the primary fracture corresponding to the first depth burst in the curve, and the other cracks were formed in the further indentation process. Based on the predication, the primary fracture location is lined out in Fig. 2.6a, and it is notable that the primary fracture occurs at the indentation location. After the ultrasonic cleaning, the cracked Cu_6Sn_5 fragments were washed away, as in Fig. 2.6c, but the surrounding Cu_6Sn_5 grains still show no breakage. Figure 2.6d, e show the top views of the target grain before and after the test. Although the primary fracture surface is not the fracture surface in Fig. 2.6e, it can be determined by comparing the front view and the top view of the target Cu_6Sn_5 grain based on a charting principle. After determining the fracture surfaces, their areas can be measured and the fracture strength of the Cu_6Sn_5 grains can be calculated.

Based on the foresaid discussions on the indentation curves and the fracture morphologies, the fracture processes of the two fracture modes are illustrated in Fig. 2.7. In all the figures, the contours of the Cu_6Sn_5 grains at the last stage are presented with the broken lines to clearly show the fracture process. Figure 2.7a1–a3 show the fracture process of the Cu_6Sn_5 grain shown in Fig. 2.6. During the initial deforming stage, the Cu_6Sn_5 grain deforms elastically, as in Fig. 2.7a1. When the load increases to a critical value, the Cu_6Sn_5 grain fractures at its center portion (see Fig. 2.7a2), inducing the first depth burst on the force–depth curve. As the fracture location is very close to the indentation location, the flexural torque on the fracture plane is little and the fracture should be induced by the shear stress. In the latter process, the indenter is sustained by the residual part of the target Cu_6Sn_5 grain, as shown in Fig. 2.7a3; some parallel secondary cracks are formed, while the surrounding Cu_6Sn_5 grains are not broken. The fracture process of the Cu_6Sn_5 grain in Fig. 2.5 is illustrated in Fig. 2.7b1–b3. As in the figures, the Cu_6Sn_5 grain also only displays elastic deformation at the first stage, but fractures at its foundation when the load increases to a certain value. Because there is an arm of force between the fracture location and the indentation location, the Cu_6Sn_5 grain fractures like a cantilever; i.e., it is the flexural torque that causes the fracture, and the fracture occurs at the foundation because the flexural torque there is the highest. The depth burst in this fracture mode is much higher than the first fracture mode, because the indenter should descend a larger distance until it can be sustained by the Cu_6Sn_5 grains around the target grain. As a result, the secondary fracture occurs in the surrounding Cu_6Sn_5 grains rather than in the residual part of the target Cu_6Sn_5 grain. It is predicated that the slender Cu_6Sn_5 grains are more likely to fracture at the foundation at a lower stress, while the podgy grains tend to fracture at the center portion at a higher stress. It has been widely accepted that the fracture

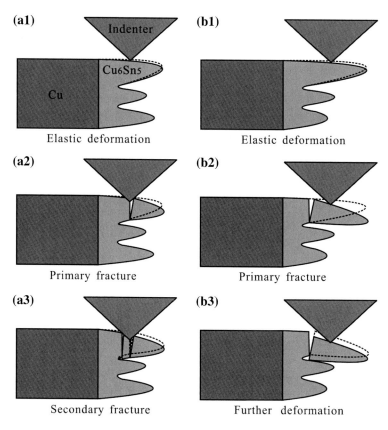

Fig. 2.7 Illustrations on fracture processes of Cu$_6$Sn$_5$ grains fractured **a1–a3** at the center portion and **b1–b3** at the foundation. Reprinted from Ref. [18]. Copyright 2011, with permission from AIP Publishing LLC

behavior of the IMCs layer is affected by their thickness and morphologies, the present work provides a strong and direct support for that.

As discussed above, the primary fracture at the center portion is induced by the shear stress; therefore, the fracture strength in this condition can be considered similar to the shear strength of the Cu$_6$Sn$_5$ grains. After measuring the fracture surfaces of the Cu$_6$Sn$_5$ grains, the shear stresses were calculated by dividing the force with the fracture area. Figure 2.8 shows the first stage of the strength–displacement curves of a few Cu$_6$Sn$_5$ grains fractured at their center portion, in which the bursts occur at the stresses as high as about 670 MPa. Since the bursts correspond to the primary shear fractures in the Cu$_6$Sn$_5$ grains, the shear fracture strength of Cu$_6$Sn$_5$ is easily estimated to be around 670 MPa. This is the first report on the shear fracture strength of the Cu$_6$Sn$_5$ intermetallic compounds. Jiang et al. [11] have reported that the Cu$_6$Sn$_5$ micropillar fractures along certain crystallographic planes under compressive loadings, the fracture stress was around

Fig. 2.8 The first stage of shear stress–displacement curves of the Cu_6Sn_5 grains fractured at their center portion. Reprinted from Ref. [18]. Copyright 2011, with permission from AIP Publishing LLC

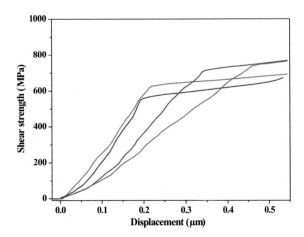

1356 MPa and the angle between the compressive direction and the fracture plane is about 60°. Based on that, the shear fracture stress at the fracture plane is estimated to be about 587 MPa through dividing the shear component of the compressive stress by the area of the fracture plane, which is close to the current results.

It is widely known that with the decreasing trend in size of the solder joints, the solder joints will contain fewer Cu_6Sn_5 grains, makes the fracture behaviors of the Cu_6Sn_5 grains more influential on the strength of the solder joints [21–23]. Therefore, the results on the shear strength of the Cu_6Sn_5 grains will be very important for evaluating the reliability of the solder joints. In addition, as the indentation test in this study is easy to carry out, it is expected that this experimental method can be popularized in investigation of the fracture strengths of the IMCs.

2.4   Fracture Behavior of Cu_6Sn_5 Induced by Deformation of Solder

2.4.1   Growth Behavior of Cu_6Sn_5 at Sn–4Ag/Cu Interface

The microstructures of the Sn–4Ag/Cu joint interfaces reflowed for different times and morphologies of the corresponding interfacial IMC grains are shown in Fig. 2.9. Figure 2.9a presents the interface reflowed for 1 min, in which the interfacial IMC layer is very thin, and was confirmed to be pure Cu_6Sn_5 by energy dispersive spectroscopy (EDS). In the corresponding images of the Cu_6Sn_5 grains shown in Fig. 2.9b, it can be found that their sizes are around 1–2 μm. For the solder joint reflowed for 3 min, the interfacial IMC layer is obviously thicker (see Fig. 2.9c), and it is still pure Cu_6Sn_5. Meanwhile, the sizes of the Cu_6Sn_5 grains become a little bit different, as in Fig. 2.9d, the grain size is in the range of 1–4 μm. After reflowed for 8 min, the IMC layer becomes much thicker, and their grain size is in the range of

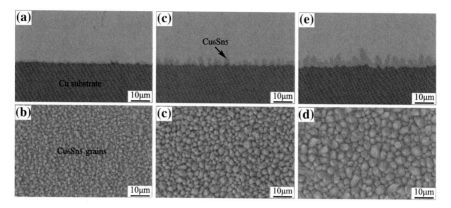

Fig. 2.9 Interfacial microstructure and morphologies of Cu$_6$Sn$_5$ grains at the Sn–4Ag/Cu interface reflowed for **a** and **b** 1 min, **c** and **d** 3 min, **e** and **f** 8 min. Reprinted from Ref. [19]. Copyright 2012, with permission from AIP Publishing LLC

2–6 μm (see Fig. 2.9e, f), implying that the difference in grain size becomes more serious with increasing reflow time. Thermodynamically, there should be a Cu$_3$Sn layer between the Cu and Cu$_6$Sn$_5$, but the Cu$_3$Sn layer is usually very thin after reflow, because its formation requires extended contact times [3, 24]. Since the SEM cannot reveal the three-dimensional (3D) morphologies of the Cu$_6$Sn$_5$ grains, the measuring laser confocal microscope was employed to observe the 3D morphologies of the Cu$_6$Sn$_5$ grains, in order to reveal their shape more visually.

The 3D morphologies of the Cu$_6$Sn$_5$ layers in Fig. 2.9 are shown in Fig. 2.10. At the joint interface reflowed for 1 min, the Cu$_6$Sn$_5$ layer is very flat, as in Fig. 2.10a, few prominent Cu$_6$Sn$_5$ grains can be observed. Figure 2.10b shows the Cu$_6$Sn$_5$ grains reflowed for 3 min, it is notable that the Cu$_6$Sn$_5$ layer becomes much coarser, the grain size is much larger and some protrudent Cu$_6$Sn$_5$ grains appear, which may be attributed to the difference in growth rates of the Cu$_6$Sn$_5$ grains with different orientations [3, 25]. After reflowed for 8 min, the increases in grain size and surface roughness of the Cu$_6$Sn$_5$ layer are more obvious, and there are much more protrudent Cu$_6$Sn$_5$ grains (see Fig. 2.10c). The images in Fig. 2.9a, c, e are actually cross sections of these Cu$_6$Sn$_5$ layers, thus they consist well with the 3D images.

To analyze the stress applied on the Cu$_6$Sn$_5$ grains and their fracture behavior, it is necessary to give a quantitative description on their shape. According to Figs. 2.9 and 2.10, the protrudent Cu$_6$Sn$_5$ grains are centrosymmetric grains with the contour lines of their cross sections similar to the parabola; thus in this study, they are approximately described by a revolution body of the parabola $y = ax^2$, in which y is the length of the Cu$_6$Sn$_5$ grain, x is the radius and a is a shape parameter. The underside radius of the Cu$_6$Sn$_5$ grain is defined as r. Since fracture of the Cu$_6$Sn$_5$ layer usually occurs in the protrudent Cu$_6$Sn$_5$ grains, the length and underside radius of the five largest Cu$_6$Sn$_5$ grains in Fig. 2.9a, c, e were measured, the average sizes are obtained as below, respectively:

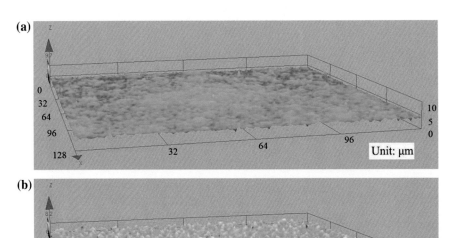

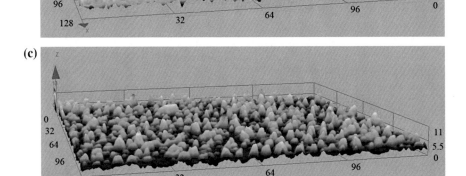

Fig. 2.10 Three-dimensional images of Cu$_6$Sn$_5$ grains at the Sn–4Ag/Cu interface reflowed for **a** 1 min, **b** 3 min and **c** 8 min. Reprinted from ref. [19], Copyright 2012, with permission from AIP Publishing LLC

$$l_{1min} = 1.52\,\mu m, \quad r_{1min} = 1.1\,\mu m$$
$$l_{3min} = 3.84\,\mu m, \quad r_{3min} = 1.67\,\mu m$$
$$l_{8min} = 6.41\,\mu m, \quad r_{8min} = 1.92\,\mu m$$

According to the underside radius and length of the Cu$_6$Sn$_5$ grains, their shape parameters were calculated to be $a_{1min} = 1.26$, $a_{3min} = 1.37$, $a_{8min} = 1.74$, respectively. In consequence, the shape of the Cu$_6$Sn$_5$ grains can be described quantitatively. The stress applied on the Cu$_6$Sn$_5$ grains will be analyzed later to express the fracture behaviors of the latter.

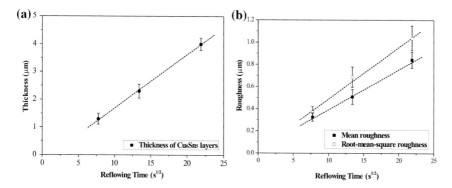

Fig. 2.11 a Interfacial Cu_6Sn_5 thickness–reflow time relationship, **b** roughness–reflow time relationship. Reprinted from Ref. [19]. Copyright 2012, with permission from AIP Publishing LLC

Figure 2.11 shows the average thickness of the Cu_6Sn_5 layers and the roughness of the solder/Cu_6Sn_5 interfaces at different solder joints. As in Fig. 2.11a, the thickness of the Cu_6Sn_5 layers increase linearly with increasing square root of the reflow time. It is about 1.2 μm at the interface reflowed for 1 min, 2.4 μm at the interface reflowed for 3 min, and 4 μm at that reflowed for 8 min. As the thicknesses of the Cu_6Sn_5 layers are close to their grain sizes, it can be confirmed that the Cu_6Sn_5 layer is a single layer of grains. Since the surface roughness of the Cu_6Sn_5 layer is a reflection of the protrudent segment of the Cu_6Sn_5 grains, they may affect the fracture behavior of the latter. Therefore, the roughness of the Cu_6Sn_5 layers were measured by measuring laser confocal microscope, both the root-mean-square roughness and the mean roughness were exported. As in Fig. 2.11b, the two roughness parameters also increase approximately linear with increasing square root of the reflow time, and their difference increases with increasing reflow time, induced by the greater difference in size of Cu_6Sn_5 grains at the long-term reflowed joint interface. Since the mean roughness equals to half of the average distance between the peaks and bottoms of the surface of the Cu_6Sn_5 layer, the average protrudent length of the Cu_6Sn_5 grains is about 1.6 μm, and that of the protrudent Cu_6Sn_5 grains can be twice as longer as that length. The differences in thickness and morphologies of the interfacial Cu_6Sn_5 grains may affect the stress applied on them and in turn their fracture behaviors.

2.4.2 Fracture Behavior at Sn–4Ag/Cu Interface

The tensile stress–strain curves and strength of the solder joints are shown in Fig. 2.12. Figure 2.12a shows the stress–strain curves of the solder joint reflowed for 3 min and tested at three different strain rates. The strains were calculated by first subtracting the elastic displacement of the Cu substrate from the total displacement, and then divided the result with the thickness of the solder in the solder

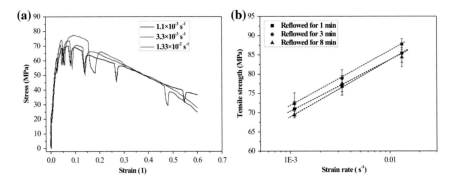

Fig. 2.12 Tensile behaviors of Sn–4Ag/Cu solder joints: **a** tensile stress–strain curves of the solder joint reflowed for 3 min; and **b** tensile strengths at three different strain rates. Reprinted from Ref. [19]. Copyright 2012, with permission from AIP Publishing LLC

joints. Therefore, the tensile curves of the solder joints are very similar to the tensile curves of the solder alloys [26, 27], the stress decreases shortly after yielding of the solder, and also the elongation is very high. The three curves have similar styles and approximately coincident elastic stages, the tensile strength (UTS) at higher strain rate is a little higher, and the corresponding strain to the UTS is around 0.06–0.08. During the holding period, obvious stress relaxation occurred. Since the Sn–Ag solder has superior ductility, the solder joints fracture at very high strain. Different from the pure solder sample, not only necking of the solder but also local fracture in the joint interface can induce the decrease in stress of the solder joint.

The average UTSs of the solder joints tested at different strain rates are exhibited in Fig. 2.12b. As in the figure, the UTS of the solder joints reflowed for the same time increase with increasing strain rate, and the UTS of the solder joints reflowed 3 min is higher than the other two groups, albeit only a little. Since the solders in all the solder joints are the same, the difference in UTSs of the solder joint reflowed for different times can only be resulted from their different interfacial microstructures. Besides, the UTSs of solder joints also show exponential increase with increasing strain rate, which is similar to the relationship between the strain rate and yield strength of the solder alloy [27, 28]. The influencing mechanisms of strain rate and reflow time will be discussed later in detail.

To reveal the interfacial fracture mechanisms, fracture processes of the solder joints were in situ observed. Figure 2.13 shows the interfacial fracture processes of the solder joints reflowed for 1 min and tested at the strain rate of 1.33×10^{-2}, with the strains labeled in each figure. At the early stage of the tensile process, the deformation is slight and only visible around the joint interface. As in Fig. 2.13a, a deformation step was formed at the Cu_6Sn_5/solder interface, which was induced by deformation mismatch between the solder and the Cu substrate. With increasing strain, the interfacial deformation becomes obvious, and serious strain concentration occurs inside the solder close to the joint interface (see Fig. 2.13b). As a result, the interfacial damage develops very fast, resulting in the initiation of some

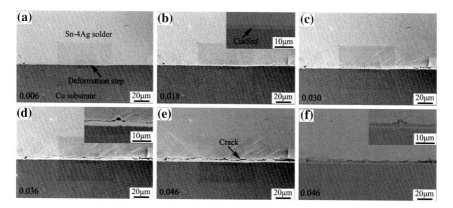

Fig. 2.13 Interfacial fracture morphologies of Sn–4Ag/Cu solder joints reflowed for 1 min and tested at the strain rate of 1.33×10^{-2} s^{-1}. Reprinted from Ref. [19]. Copyright 2012, with permission from AIP Publishing LLC

microcracks at the interface, as in Fig. 2.13c. In the further deformation process, the microcracks become wider and gradually connect with each other, inducing the interfacial fracture, as in Fig. 2.13d, e. In the backscattered electron image (Fig. 2.13f), it can be found that fracture occurs in the solder around the Cu_6Sn_5/solder interface, and there is no cracking in the Cu_6Sn_5 layers. For the joint interface reflowed for 1 min, the strain concentration zone is thin and the interfacial fracture occurs at a low strain.

The interfacial fracture process of the solder joints reflowed for 3 min and tested at two different strain rates are exhibited in Fig. 2.14; the stains were also labeled in the figures. Figure 2.14a–c show the interfacial fracture process of the solder joints

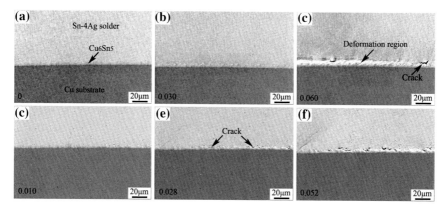

Fig. 2.14 Interfacial fracture morphologies of Sn–4Ag/Cu solder joints reflowed for 3 min and tested at the strain rate of **a–c** 3.3×10^{-3} s^{-1} and **d–f** 1.33×10^{-2} s^{-1}. Reprinted from Ref. [19]. Copyright 2012, with permission from AIP Publishing LLC

tested at the strain rate of 3.3×10^{-3} s^{-1}. As in Fig. 2.14a, b, obvious deformation
mismatch between the solder and the Cu_6Sn_5 layer occurs shortly after the tensile
test. At higher strain, deformation of the solder close to the joints interface becomes
serious, and some microcracks appear inside the solder, as in Fig. 2.14c. Compared
with the solder joints reflowed for 1 min, the strain concentration zone is a bit
wider, and there is also no fracture in the Cu_6Sn_5 layer. Figure 2.14d–f show the
fracture process of the solder joints tested at the strain rate of 1.33×10^{-2} s^{-1}. As in
the figures, deformation of the solder also increases with increasing strain, and the
step induced by the deformation mismatch is more obvious, but the plastic deformation is less serious since the yield strength of the solder is higher at higher strain
rate. Besides, some Cu_6Sn_5 grains fractured during the tensile process, indicating
that they are more apt to fracture at higher strain rate.

Figure 2.15 shows the interfacial fracture process of the solder joints reflowed
for 8 min and tested at the strain rate of 1.1×10^{-3}. Before the tensile test, the side
surface of the sample is very flat, as in Fig. 2.15a. At the strain of 0.02, deformation
mismatch started to emerge at interface (see Fig. 2.15b). In Fig. 2.15c, d, some
Cu_6Sn_5 grains have fractured, although plastic deformation of the solder is not very
serious. During the latter deforming process, the cracks in the Cu_6Sn_5 grains
propagated into the solder and connected with each other, inducing the fracture
along the joint interface, as in Fig. 2.15e, f. According to Figs. 2.13, 2.14, and 2.15,
it can be concluded that there are always deformation mismatch and strain concentration inside the solder around the Cu_6Sn_5/solder interface. The solder joints
always fracture around the joint interface, but the exact fracture location is a little
bit different and affected by the reflow times and tensile strain rate. The solder joints
reflowed for a long time or deformed at higher strain rate are more apt to fracture
inside the IMC layer.

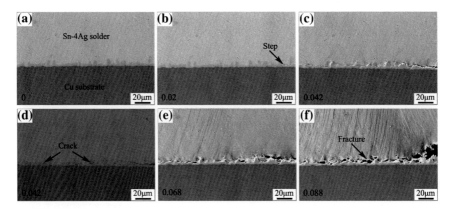

Fig. 2.15 Interfacial fracture morphologies of Sn–4Ag/Cu solder joints reflowed for 8 min and
tested at the strain rate of 1.1×10^{-3} s^{-1}. Reprinted from Ref. [19]. Copyright 2012, with
permission from AIP Publishing LLC

2.4.3   Applied Stress and Fracture Behavior of Cu$_6$Sn$_5$ Grains

To further reveal the fracture behaviors of the Cu$_6$Sn$_5$ grains, the stress applied on them are quantitatively analyzed. As the stress endured by the Cu$_6$Sn$_5$ grain is applied by the solder, the shear stress is calculated through dividing the shear force with the underside area of the grain, and the shear force is obtained by multiplying the stress of the solder with the projection area of the hood face of the Cu$_6$Sn$_5$ grain. Since the shape of the grain is described by the revolving body of the parabola $y = a \cdot x^2$, the force endured by the Cu$_6$Sn$_5$ grain is calculated by the follow equation:

$$F = 2\tau_{solder} \int_0^{r^2} 2x dy = 4\tau_{solder} \int_0^r x da x^2 = a\tau_{solder} \frac{8}{3} r^3 \qquad (2.1)$$

and the stress is calculated to be:

$$\tau_{IMC} = F / \pi r^2 = \frac{8a\tau_{solder} r^3}{3\pi r^2} = 0.849 ar\tau_{solder} \qquad (2.2)$$

where F is the shear force, τ_{solder} is the stress in the solder close to the Cu$_6$Sn$_5$/solder interface which envelops Cu$_6$Sn$_5$ grain, and τ_{IMC} is the shear stress at the foundation of the Cu$_6$Sn$_5$ grain; the size r is a dimensionless number in this equation.

Based on Eq. (2.2) and the shape of the Cu$_6$Sn$_5$ grains exhibited in Sect. 2.4.1, the stresses endured by the protrudent Cu$_6$Sn$_5$ grains at the joint interfaces reflowed for different times were calculated and are shown Fig. 2.16. It can be found that τ_{IMC} increases obviously with increasing reflow time. For the interface reflowed for 8 min, the shear stress on the protrudent Cu$_6$Sn$_5$ grains is 2.836 times as large as that in the solder. The von Mises stress distribution inside the solder joint has also revealed that the stress at the corner of the solder is about twice as high as the stress in the solder joint [13]. Since the tensile stress in this study is about 70 MPa, the shear stress on the protrudent Cu$_6$Sn$_5$ grains was estimated to be about 300–400 MPa, which is close to their fracture stress under shear loading and bending moment [18]. Therefore, the protrudent Cu$_6$Sn$_5$ grains at these interfaces usually fracture during the tensile or shear process [29]. In contrast, both the shear stress and the bending moment on the small Cu$_6$Sn$_5$ grains are much lower, thus they can hardly fracture prior to fracture of the solder.

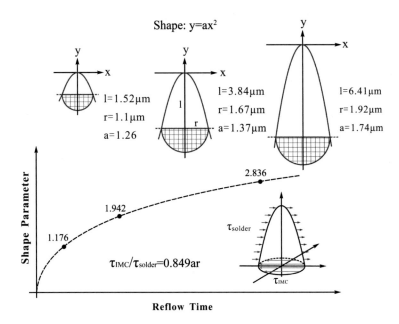

Fig. 2.16 Analysis diagram of the shear stress applied on the Cu_6Sn_5 grains reflowed for different times. Reprinted from Ref. [19]. Copyright 2012, with permission from AIP Publishing LLC

2.5 Fracture Behavior of Cu–Sn IMCs Induced by Deformation of Substrate

2.5.1 Morphology of IMCs at Sn–4Ag/Cu Interface

The morphologies of the IMC layers at the as-soldered and aged Sn–4Ag/Cu interfaces were observed and are shown in Fig. 2.17. Figure 2.17a presents the morphology of the as-soldered Sn–4Ag/Cu single crystal interface, in which the IMC thickness is about 2 μm, and Energy Dispersive X-ray Detector (EDX) analysis indicates that it is Cu_6Sn_5. After aging for 4 days, the IMC thickness increases to about 7 μm, as in Fig. 2.17b, and a fuscous IMC layer appears between the Cu and Cu_6Sn_5 layer, which was confirmed to be Cu_3Sn by EDX. The morphologies of the Sn–Ag/cold-drawn Cu interfaces are a little bit different from that of the Sn–Ag/Cu single crystal interfaces. As in Fig. 2.17c, the IMC layer at the as-soldered Sn–Ag/cold-drawn Cu interface is a little thicker, and some protrudent Cu_6Sn_5 grains were observed. That may attribute to the difference in microstructure of the Cu substrates, because the growth rates of the IMC grains at different crystallographic planes of the Cu crystal are different [24, 25]. Since the cold-drawn Cu is composed of thin crystal grains with different orientations, the Cu_6Sn_5 grains at different grains are different in growth rate, making some of them become protrudent. After aging for 4 days, the IMC layer becomes much thicker,

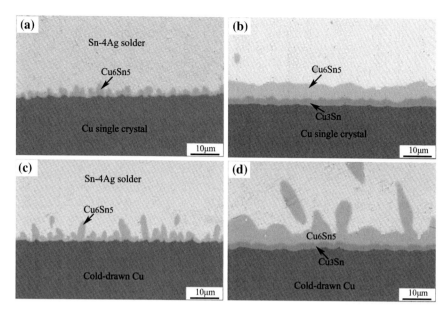

Fig. 2.17 Interfacial morphologies of Sn–4Ag/Cu single crystal solder joints **a** as-reflowed and **b** aged for 4 days; Sn–4Ag/cold-drawn Cu solder joints **c** as-reflowed and **d** aged for 4 days. Reprinted from Ref. [20]. Copyright 2011, with permission from Elsevier

and some protrudent Cu$_6$Sn$_5$ grains still exist (see Fig. 2.17d). In general, the IMC layers at the Sn–Ag/cold-drawn Cu interface are a little thicker and coarser than that at the Sn–4Ag/Cu single crystal interface.

2.5.2   Fracture Behavior of Interfacial IMC Layer

The tensile stress–strain curves of the two groups of Sn–Ag/Cu specimens are shown in Fig. 2.18. As the IMC layers are very thin and the solder is very soft, the curves can be approximately regarded as the tensile curves of the Cu substrates. Figure 2.18a shows the stress–strain curves of the Sn–4Ag/Cu single crystal sample, it can be seen that the yield strength is about 32 MPa, and there is a long strain hardening stage after yielding. Even when the strain has exceeded 6 %, the specimen shows no necking or fracture. In contrast, the yield strength of the cold-drawn Cu is much higher, while the strain hardening stage is very short. As in Fig. 2.18b, the yield strength of the cold-drawn Cu is about 300 MPa and the yield strain is about 1 %, and necking occurs shortly after yielding. When the tests were paused at certain strains, the stress shows slight decrease, which was induced by stress relaxation. The aged Sn–Ag/cold-drawn Cu sample exhibits a lower yield strength but better ductility because recovery occurs during the aging process. In contrast, since there are few defects in the Cu single crystal, the thermal aging has

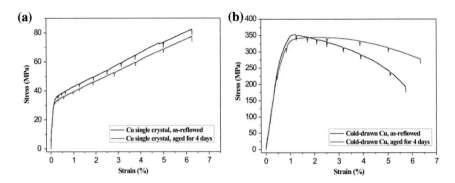

Fig. 2.18 Nominal stress–strain curves of samples: **a** Sn–Ag/Cu single crystal samples, **b** Sn–Ag/cold-drawn Cu samples. Reprinted from Ref. [20]. Copyright 2011, with permission from Elsevier

little influence on it, and the difference in their yield strength was predicated to be induced by their different Schmidt factors. The deformation behavior of the Cu substrates is dominative on fracture behaviors of the interfacial IMC layers, which will be shown later in detail.

The fracture morphologies of the IMC layer (Cu_6Sn_5) at the as-soldered Sn–Ag/Cu single crystal interface are shown in Fig. 2.19, and the strains are tagged in each figure. Figure 2.19a exhibits the interfacial morphology at a strain of 6.3×10^{-3}, in which obvious slip bands (SBs) have appeared and a few cracks inside the IMC layer are visible. In comparison, the thin Cu_6Sn_5 layer in the as-soldered tensile solder joints can hardly fracture [13, 26]. Besides, there is clear correspondence between the cracks and the slip bands, which is similar to the Sn–Ag–Cu/Cu interface deformed under fatigue loadings [30]. However, there is no fracture in the IMC layer before the slip bands emerge and act on the IMC/Cu

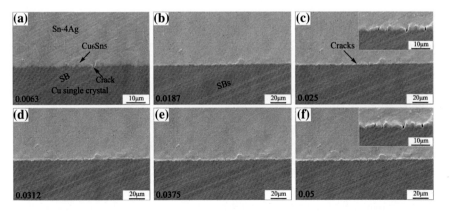

Fig. 2.19 Fracture behaviors of as-soldered Sn–Ag/Cu single crystal interfaces at different strains: **a** $\varepsilon = 6.3 \times 10^{-3}$, **b** $\varepsilon = 1.87 \times 10^{-2}$, **c** $\varepsilon = 2.5 \times 10^{-2}$, **d** $\varepsilon = 3.12 \times 10^{-2}$, **e** $\varepsilon = 3.75 \times 10^{-2}$, **f** $\varepsilon = 5 \times 10^{-2}$. Reprinted from Ref. [20]. Copyright 2011, with permission from Elsevier

interface. With increasing strain, the slip bands become more and more serious, as in Fig. 2.19b, c, and the width and number of the cracks increase obviously. In the latter deforming process, only the width of the cracks keeps increasing, while their number shows little increase (see Fig. 2.19d–f). According to the correspondence between the slip bands and the cracks in the IMC layer, it is deduced that the cracks may be induced by impingement of the slip bands in the Cu substrate.

Figure 2.20 shows the fracture morphologies of the IMC layer ($Cu_6Sn_5 + Cu_3Sn$) at the Sn–Ag/Cu single crystal interface aged for 4 days, and the strains are also tagged in each figure. In Fig. 2.20a, the strain is only 1.56×10^{-3}, but a few thin cracks have been observed in the IMC layer. However, there is also no cracking in the IMC layer before the slip bands act on the IMC/Cu interface. In Fig. 2.20b, c, both the number and the width of the cracks increase obviously, and the corresponding relationship between the slip bands and the cracks is more obvious. Although the Cu_6Sn_5 and Cu_3Sn layers are a little bit different in mechanical properties [5], the vertical cracks can easily propagate from the Cu_3Sn into Cu_6Sn_5, because the Cu_6Sn_5 and Cu_3Sn are hard and brittle and are closely bonded. In contrast, the cracks cannot get through the Cu_6Sn_5/solder interface. In Fig. 2.20d–f, the evolution of the cracks is similar to that in Fig. 2.19, i.e., the width of the cracks increases obviously but no further increase in the number. Generally, the fracture behaviors of the IMC layer in Figs. 2.19 and 2.20 are quite different from that in the solder joint under tensile or shear loading [13, 26, 31]. There is no strain concentration inside the solder close to the joint interface, and the cracks are vertical to the joint interface. In fact, since the Cu substrate is the carrier, the deformation of the IMC layer is driven by the substrate, and then the solder is driven to deform by the IMC layer. Therefore, the solder can deform freely and no restraint stress is applied on the IMC layer to make it fracture. Considering the correspondence between the slip bands and the vertical cracks, it is concluded that it is the impingement of the slip bands (dislocations) that induces the fracture in the IMC layer.

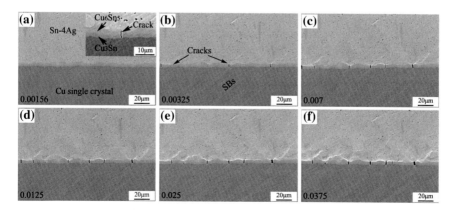

Fig. 2.20 Fracture behaviors of aged Sn–Ag/Cu single crystal interfaces at different strains: **a** $\varepsilon = 1.56 \times 10^{-3}$, **b** $\varepsilon = 3.25 \times 10^{-3}$, **c** $\varepsilon = 7 \times 10^{-3}$, **d** $\varepsilon = 1.25 \times 10^{-2}$, **e** $\varepsilon = 2.5 \times 10^{-2}$, **f** $\varepsilon = 3.75 \times 10^{-2}$. Reprinted from Ref. [20]. Copyright 2011, with permission from Elsevier

Fig. 2.21 Fracture morphologies of IMC grains at Sn–Ag/Cu single crystal interfaces: **a** and **b** as-reflowed samples; **c** and **d** samples aged for 4 days. Reprinted from Ref. [20]. Copyright 2011, with permission from Elsevier

The morphologies of the IMC grains at the Sn–Ag/Cu single crystal interfaces after the tests are shown in Fig. 2.21, in which the loading direction is indicated by the arrows. The observed IMCs are Cu_6Sn_5 grains as the Cu_3Sn is covered by them. Figure 2.21a, b presents the morphologies of the as-soldered Cu_6Sn_5 grains; it was found that the grains are not very compact and their grain size is about a few microns. The cracks are similar in direction, and both transgranular and intergranular cracks appear. The Cu_6Sn_5 grains at the aged interfaces are much larger, compact, and equiaxed, and the cracks are more regular. As in Fig. 2.21c, the cracks are straight and parallel with each other. Because the cracks are induced by the slip bands, they have similar distribution with the latter, and the widths are different as they fracture at different time. At higher magnification, the transgranular fracture is found to be the major fracture mechanisms, although fracture may also occur along the grain boundaries when the slip bands are close to them (see Fig. 2.21d). Compared with the cracks in the as-soldered IMC layer, the cracks in the aged IMC layer are much wider but their density is lower, which is well consistent with the interfacial fracture morphologies in Figs. 2.19 and 2.20.

The fracture behaviors of the IMC layers at the Sn–Ag/cold-drawn Cu interfaces are a little bit different from that at the Sn–Ag/Cu single crystal interfaces, as shown in Figs. 2.22 and 2.23. Figure 2.22a presents the morphology of the as-soldered Sn–Ag/cold-drawn Cu interface at a strain of 1.24×10^{-2}, in which the sample just started to yield, some slip bands appeared at the surface of the Cu substrate and a few cracks in the IMC layer (pure Cu_6Sn_5) are visible. Compared with the as-reflowed Sn–Ag/Cu single crystal interface, the critical "fracture strain" of the

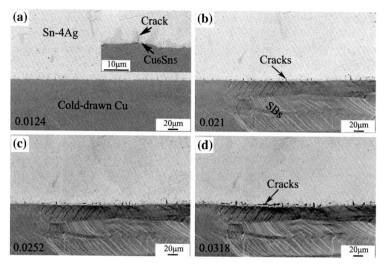

Fig. 2.22 Fracture behaviors of as-reflowed Sn–Ag/cold-drawn Cu interfaces at different strain: **a** $\varepsilon = 1.24 \times 10^{-2}$, **b** $\varepsilon = 2.1 \times 10^{-2}$, **c** $\varepsilon = 2.52 \times 10^{-2}$, **d** $\varepsilon = 3.18 \times 10^{-2}$. Reprinted from Ref. [20]. Copyright 2011, with permission from Elsevier

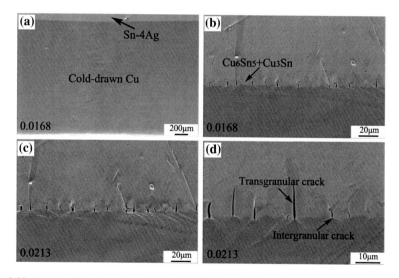

Fig. 2.23 Fracture behaviors of aged Sn–Ag/cold-drawn Cu interfaces: **a** macroscopic deformation and **b** microscopic fracture morphologies of joint interface at the strain of 1.68×10^{-2}; **c** and **d** fracture morphologies of joint interface at the strain of 2.13×10^{-2}. Reprinted from Ref. [20]. Copyright 2011, with permission from Elsevier

IMC layer (the strain for occurrence of the fracture in the IMC layer) is much higher. However, there is also no fracture in the IMC layer before the Cu substrate yields. The necking of the cold-drawn Cu sample occurs shortly after the yielding and generates a serious strain concentration. At higher magnification, some cracks are found to be very sharp, indicating that they initiate at the IMC/Cu interface and propagate to the IMC/solder interface. Figure 2.22b shows the necking region of the specimen at a strain of 2.1×10^{-2}, the cracks in the IMC layer are serious, and there is also clear correspondence between the cracks and the slip bands. Whereas, the plastic deformation of the Cu substrate beyond the necking region is not obvious, and there is no cracking in the IMC layer. During the further deforming process, the cracks become wider, gradually turn to the solder/IMC interface and connect with each other to form long cracks (see Fig. 2.22c, d). That phenomenon does not appear in Figs. 2.19 and 2.20 because the deformation of Cu substrates is not so serious. Due to the serious strain concentration, deformation and fracture at the necking region develop very fast.

The fracture behaviors of the IMC layer (Cu_6Sn_5 + Cu_3Sn) at the aged Sn–Ag/cold-drawn Cu interface are similar to that at the as-reflowed interface, and the critical "fracture strain" of the IMC layer is about 1.1×10^{-2}, as exhibited in Fig. 2.23. Figure 2.23a shows the macroscopic deformation morphology at a strain of 1.68×10^{-2}, it is notable that the plastic deformation is obvious at the necking region, but slight at the other region. In microscale, the cracks were only observed at the necking region, and there is also clear correlation between the slip bands and the vertical cracks, as in Fig. 2.23b. Compared with that shown in Fig. 2.20, the plastic deformation is complex and the slip bands are different in orientation, making the cracks in the IMC layer a bit random. In Fig. 2.23c, d, the slip bands and the cracks become wider but their number changes little, both the intergranular and the transgranular cracks can be observed at higher magnification, and some sharp cracks are obvious. In general, the IMC layers in Figs. 2.22 and 2.23 fracture at high stress/strain, but that only occurs at the necking region, and there is also correspondence between the slip bands and the cracks in the IMC layers.

The morphologies of the cracked IMC grains at the Sn–Ag/cold-drawn Cu interfaces are shown in Fig. 2.24, the loading direction is also indicated by the arrows. Figure 2.24a, b display the Cu_6Sn_5 grains at the as-reflowed interface. The cracks are approximately parallel and vertical to the loading direction, and can be either intergranular or transgranular. Although the two figures are taken from the same sample, the widths of the cracks are quite different since the plastic strains at different locations are different. The Cu_6Sn_5 grains at the aged interface are larger and equiaxed, and the intergranular and the transgranular cracks are clearer, as in Fig. 2.24c, d. Besides, the cracks are not exactly parallel with each other and the intergranular cracks are more common, because the cold-drawn Cu is composed of thin grains with different orientations and accordingly random slip bands, making the cracks in the IMC layer less regular.

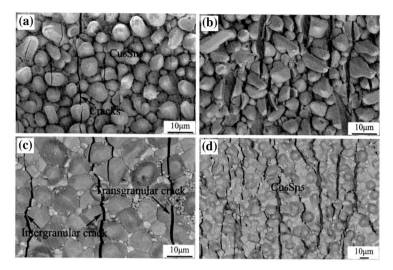

Fig. 2.24 Fracture morphologies of IMC grains at Sn–4Ag/cold-drawn Cu interfaces: **a** and **b** as-reflowed samples; **c** and **d** samples aged for 4 days. Reprinted from Ref. [20]. Copyright 2011, with permission from Elsevier

2.5.3   Fracture Mechanisms Induced by Deformation of Substrate

According to the observations above, there is definite correspondence between the slip bands and the cracks in the IMC layers. Besides, the fracture process of the IMC layer is similar to the transition of plastic deformation between the metallic grains. In the polycrystalline metals, the slip systems of the grains with preferential orientations are stimulated firstly and the dislocations in these grains slip outwards to the grain boundary [32, 33]. As the dislocations cannot get across the larger angle grain boundaries, they pile-up at the boundaries and induce a cumulative stress ahead the pile-up group [32, 34]. When the number of the piled dislocations increases to a critical value, the cumulative stress will stimulate the slip system in the adjacent grain [35, 36], or sometimes induces a microcrack [32]. As the fracture in the IMC layer occurs when the dislocations impinge on it, it may be also induced by the piled dislocations. It has been well accepted that the Cu_6Sn_5 has two structural forms, i.e. the conventional NiAs-type structure (η) and the ordered long-period superlattice structure (LPS) based on the NiAs-type structure (η') [3, 37], and Cu_3Sn has a long-period superstructure (ε) [38], both of them are essentially different from the face-centered cubic Cu substrates in lattice structure. Therefore, the dislocations in the Cu substrates cannot get across the Cu_6Sn_5/Cu or Cu_3Sn/Cu interfaces, and it is probable for them to pile-up at the two interfaces and induces the fracture.

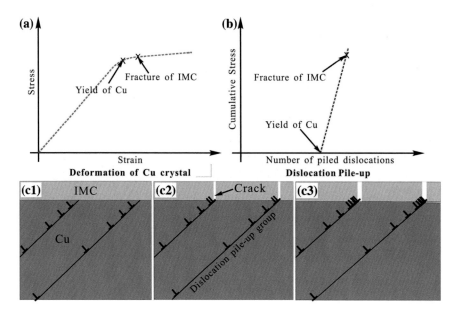

Fig. 2.25 Illustration on fracture of interfacial IMC layer induced by dislocation pile-up: **a** sketch deformation curve of a Cu crystal grain; **b** relationship between cumulative stress and piled dislocations; **c1–c3** dislocation pile-up at the interface and fracture processes of the IMC. Reprinted from Ref. [20]. Copyright 2011, with permission from Elsevier

A qualitative illustration on the fracture behaviors of the IMC layer based on the dislocation pile-up mechanism is exhibited in Fig. 2.25. Figure 2.25a shows a sketch deformation curve of a Cu grain at the IMC/Cu interface. At the beginning of the deformation process, the Cu grain deforms elastically and no fracture occurs in the IMC layer. After the Cu yields, the dislocations in the Cu grain are stimulated to slip. If there is no barrier, they will emerge at the surface and form some slip steps. However, as they cannot get across the IMC/Cu interface, they have to pile-up at the interface and induce a cumulative stress field ahead the piled-up dislocations, as shown in Fig. 2.25c1. With increasing strain, the dislocation sources in the Cu grain keep giving out dislocations, and the number of the piled dislocations and the cumulative stress increase gradually (see Fig. 2.25b). When the cumulative stress reaches the fracture strength of the IMC layer, microscale fracture occurs in the Cu_6Sn_5 or Cu_3Sn at the IMC/Cu interface. Since the IMCs are hard and brittle [13, 24], the microcracks propagate rapidly, forming the cracks vertical to the tensile direction, as in Fig. 2.25c2. In the latter process, the dislocations at the cracks increase continually with increasing strain, making the slip bands and the cracks become wider (see Fig. 2.25c3). As the fracture of the IMC layer is induced by the dislocation pile-up, yielding of the Cu substrates is a necessary condition of that.

Since the IMC/Cu interface is only at one side of the Sn–Ag/Cu specimen, the dislocation pile-up at the IMC/Cu interface is single-side style. For the single-side dislocation pile-up, the number of dislocations in the piled-up group and the cumulative stress can be estimated by the following equations [39]:

$$n = \pi(1 - v)L\tau/Gb \tag{2.3}$$

$$\tau_h = n\tau \tag{2.4}$$

where b is Burgers vector, L is the length of the dislocation pile-up groups, which approximates to the grain size of the Cu substrate, v is the Poisson ratio, G is the shear modulus, τ is the shear component of the tensile stress and τ_h is the cumulative stress on the obstacle (equal to the stress applied on the first dislocation in the pile-up group). For the Cu crystal, b is 1.81×10^{-10} m, v is 0.343 and G is 48 GPa. In Fig. 2.20, the orientation of the loading direction was determined to be around [11 20 24], accordingly the stimulated slip system is ($\bar{1}$ 1 1) [1 0 1] and the Schmidt factor is calculated to be 0.430. As the IMC layer fractures at 32 MPa, τ is calculated to be 13.76 MPa. The length of the dislocation pile-up group (L) is obtained to be 4 mm through dividing the width of the specimen with the cosine of the angle between the loading direction and slip direction. Based on that, n is calculated to be 1.31×10^4 and τ_h is 179.8 GPa in theory. However, as the hardness of Cu_6Sn_5 is about 3 GPa and its compression strength is about 1.4 GPa [11, 13], the cumulative stress can not really achieve the foreside value, because a much smaller number of piled dislocations will make the IMC layer fracture to release the cumulative stress. For the cold-drawn Cu, L is around 50–100 μm, while the shear stress (τ) is much higher, and the theoretical cumulative stress has the same order of magnitude with that at the Sn–Ag/Cu single crystal interface. Since the cumulative stress of a small group of dislocations is very high, it will soon reach the fracture strength of IMC layer after the dislocations arrive at the Cu/IMC interface, thus the IMC layers at both the two Sn–Ag/Cu interfaces fracture very shortly after yielding of the Cu substrates.

As exhibited before, the critical "fracture strain" of the IMC layer at the Sn–Ag/cold-drawn Cu interface is higher than 1 %. As its elastic modulus is about 180 GPa [13], the IMC layer should endure a stress higher than 1.8 GPa if it is continuous, which is even higher than its compressive strength [11]. Therefore, the IMC layer may not be a continuous, and the fracture proposed above are not the only mechanism to accommodate the deformation. There may be some gaps between the IMC grains that can be opened under tensile stress. To confirm that, a thin sheet was sliced from the undeformed as-soldered Sn–Ag/cold-drawn Cu joint, and ground manually to a few microns. Figure 2.26 shows its morphologies observed by SEM. As in the figure, there are thin gaps between the Cu_6Sn_5 grains, and most of them extend to the Cu substrate. Though the sample may undergo a slight plastic deformation when it was ground, there are no slip bands or transgranular fracture in the IMC layer, thus the openings can only be induced by the surface deformation, which provides direct evidence for the predication on the gaps and openings. After the aging process, the gaps may still exist, at least the IMC grain boundaries are weak links, because the IMC layer in Fig. 2.23 also fractures at

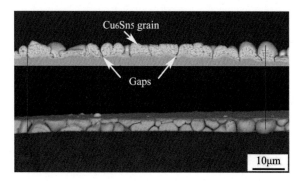

Fig. 2.26 Gaps between the Cu_6Sn_5 grains at the as-reflowed Sn–4Ag/cold-drawn Cu interface, the lines indicate the correspondence between the cross-section and the side section views. Reprinted from Ref. [20]. Copyright 2011, with permission from Elsevier

a strain higher than 1 %. There should also be such gaps between the Cu_6Sn_5 grains at the Sn–Ag/Cu single crystal interface, but the opening is less preferred because the IMC layer fractures at a low strain. The difference between the openings and the intergranular cracks is that the latter is wider and induced by dislocation pile-up. The openings are predicated to be the major mechanisms to release the surface deformation and elastic deformation, but they are not obvious and will not sharply decrease the adhesive properties of the joint interface.

The qualitative illustrations on the fracture behaviors of the IMC layers at the two Sn–Ag/Cu interfaces are shown in Fig. 2.27, with the slip bands represented by the broken lines. For the Sn–Ag/Cu single crystal interface, the fracture behavior is simple, as shown in Fig. 2.27a. After the Cu substrate yields, the dislocations slip and pile-up at the IMC/Cu interface, leading to the fracture in the IMC layer shortly after that. The cracks can be transgranular or intergranular, depending on the location of the slip bands. Similar to the distribution of the slip bands in the Cu single crystal, they are also exactly parallel to each other. Besides, there are also thin openings between the IMC grains. Figure 2.27b shows the fracture behaviors at the Sn–Ag/cold-drawn Cu interface. As the cold-drawn Cu is polycrystalline composed of grains with the size 50–100 μm and each of them has different orientations, fracture behaviors of the IMC layer are a bit complex. According to the Hall–Petch relationship, the metals with thinner grains have higher yield strength because fewer dislocations piled-up at the grain boundaries at the same stress [40]. Therefore, the slip bands in the cold-drawn Cu and the cracks in the IMC layer are thinner compared with the coarse grains. Besides, since yielding of the Cu substrate is a necessary condition for fracture in the IMC layer, the critical fracture strain of the IMC layer at the Sn–Ag/cold-drawn Cu interface is much higher, and the cracks are a bit in disorder due to the complex slip bands in the cold-drawn Cu grains. There are also both intergranular and transgranular cracks, and the intergranular cracks are more common because the slip bands are denser and have higher possibility to appear at the IMC grain boundaries. In addition, the

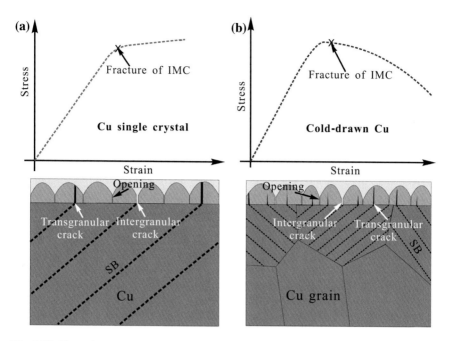

Fig. 2.27 Illustration on fracture behaviors of IMC layers at **a** Sn–Ag/Cu single crystal sample and **b** Sn–Ag/cold-drawn Cu sample. Reprinted from Ref. [20]. Copyright 2011, with permission from Elsevier

openings between the IMC grains are more serious as the surface deformation of Cu substrate is much higher.

Based on the understandings on the fracture mechanisms, some intrinsic and extrinsic factors can affect the fracture of IMC layer induced by deformation of the Cu substrate. As the IMC layer fractures when the piled-up dislocations act on it, it is very sensitive to plastic deformation of the Cu substrate. Therefore, the yield strength/strain of the Cu substrate has dominative influence on fracture behavior of the IMC layer. Because the yield strength of the Cu substrate relies mainly on its grain size, the latter can dominate the fracture behaviors of the interfacial IMC layer. The IMC layer at the Cu substrates with thin grains fractures at a high stress/strain, while that at the coarse-grained substrate fractures at a low stress. Besides, as the fracture of the IMC layer at the necking region is much more serious, it is expected that a strain concentration in the Cu substrate can induce a serious local strain, making the fracture of the IMC layer at that region more serious.

Compared with their sensitivity to the plastic deformation, the IMC layers can withstand a very high elastic deformation. According to Figs. 2.22 and 2.23, even when the strain increases to 1 %, no cracking occurs inside the IMC layer if no slip band emerges in the Cu substrate, because the gaps between the IMC grains can accommodate the elastic strain. The thermal aging has significant influence on fracture behaviors of the IMC layer as its fracture strength decreases after thermal aging [41], making it easier to fracture. Moreover, the IMC grains become much more compacted after the aging process, and the opening at the grain boundaries not so easy to occur, thus the intergranular cracking will be more favored.

2.6 Brief Summary

In this chapter, fracture behavior of the IMC layer at the Sn–4Ag/Cu solder joint interface were investigated. The shear fracture behavior and fracture strength of the Cu_6Sn_5 grain was studied using Dynamic Ultramicroscopic Hardness tester, the fracture behavior of IMC layer induced by deformation of solder and Cu substrate were in situ observed, the influencing factors on the fracture behavior were discussed. The major conclusions are as follows:

(1) Cleavage fracture occurs in the Cu_6Sn_5 grains when the shear stress applied on it increase to a certain value, result in a burst increase of strain on the indentation curves. The Cu_6Sn_5 grains usually fracture at the foundation or the center portion, depending on their size and shape. The slender Cu_6Sn_5 grains are more likely to fracture at the foundation at a lower stress, induced by shear stress and the flexural torque, while the podgy grains tend to fracture at the center portion, result from the shear stress. The fracture strength of the Cu_6Sn_5 grains fractured at their center portion is close to the shear strength of the Cu_6Sn_5 IMC, which is about 670 MPa.

(2) The thickness and surface roughness of the interfacial Cu_6Sn_5 layer at the Sn–4Ag/Cu joint interfaces increase linearly with increasing square root of the reflow time, some protrudent Cu_6Sn_5 grains appear at the joint interface after reflowed for a long time, and their shapes can be approximately described by the revolution body of parabola. Serious strain concentration occurs around the joint interface during the tensile process, the joint interfaces reflowed for longer time usually fractures in the Cu_6Sn_5 layer. For the joint interface reflowed for 8 min, the shear stress applied on the protrudent Cu_6Sn_5 grains by solder is calculated to be about 300–400 MPa. At high strain rate, the stress applied on the Cu_6Sn_5 by the solder is higher which makes the IMC layer easier to fracture.

(3) The IMC layers at the Sn–Ag/Cu single crystal interfaces and Sn–Ag/cold-drawn Cu interfaces fracture shortly after the yield of the Cu substrate, forms the cracks vertical to the joint interface which correlate well with the slip bands in the Cu substrates. In microscale, the fracture is induced by dislocation pile-up. After the Cu substrates yield, the dislocations pile-up at the IMC/Cu interface and generate a high cumulative stress ahead the pile-up group. Fractures inside the IMC layer occur when the cumulative stress reaches the fracture strength of the IMC. As a small group of piled dislocations can induce a very high cumulative stress, fracture of the IMC

layer occurs very shortly after the Cu substrates yield. Plastic deformation of the Cu substrates can be regarded as the sufficient condition for fracture of the IMC layer, and the grain size of the Cu substrate has dominative influence on the fracture behaviors. Thermal aging makes the IMC layers thicker, more compact, and easier to fracture. There are thin gaps between the IMC grains that can be opened under tensile stress to accommodate the surface deformation, thus the IMC layer can withstand a large elastic strain before the yield of the Cu substrate.

References

1. Chan YC, Yang D. Failure mechanisms of solder interconnects under current stressing in advanced electronic packages. Prog Mater Sci. 2010;55:428–75.
2. Prakash KH, Sritharan T. Interface reaction between copper and molten tin-lead solders. Acta Mater. 2001;49:2481–9.
3. Laurila T, Vuorinen V, Kivilahti JK. Interfacial reactions between lead-free solders and common base materials. Mater Sci Eng R. 2005;49:1–60.
4. Dao M, Chollacoop N, Van Vliet KJ, Venkatesh TA, Suresh S. Computational modeling of the forward and reverse problems in instrumented sharp indentation. Acta Mater. 2001;49:3899–918.
5. Deng X, Chawla N, Chawla KK, Koopman M. Deformation behavior of (Cu, Ag)–Sn intermetallics by nanoindentation. Acta Mater. 2004;52:4291–303.
6. He M, Chen Z, Qi GJ. Solid state interfacial reaction of Sn–37Pb and Sn–3.5Ag solders with Ni–P under bump metallization. Acta Mater. 2004;52:2047–56.
7. Chromik RR, Vinci RP, Allen SL, Notis MR. Nanoindentation measurements on Cu–Sn and Ag–Sn intermetallics formed in Pb-free solder joints. J Mater Res. 2003;18:2251–61.
8. Ghosh G. Elastic properties, hardness, and indentation fracture toughness of intermetallics relevant to electronic packaging. J Mater Res. 2004;19:1439–54.
9. Jang GY, Lee JW, Duh JG. The nanoindentation characteristics of Cu_6Sn_5, Cu_3Sn, and Ni_3Sn_4 intermetallic compounds in the solder bump. J Electron Mater. 2004;33:1103–10.
10. Yang PF, Lai YS, Jian SR, Chen J, Chen RS. Nanoindentation identifications of mechanical properties of Cu_6Sn_5, Cu_3Sn, and Ni_3Sn_4 intermetallic compounds derived by diffusion couples. Mater Sci Eng A. 2008;485:305–10.
11. Jiang L, Chawla N. Mechanical properties of Cu_6Sn_5 intermetallic by micropillar compression testing. Script Mater. 2010;63:480–3.
12. Kikuchi S, Nishimura M, Suetsugu K, Ikari T, Matsushige K. Strength of bonding interface in lead-free Sn alloy solders. Mater Sci Eng A. 2001;319–321:475–9.
13. Lee HT, Chen MH, Jao HM, Liao TL. Influence of interfacial intermetallic compound on fracture behavior of solder joints. Mater Sci Eng A. 2003;358:134–41.
14. Kima KS, Huh SH, Suganuma K. E ffects of intermetallic compounds on properties of Sn–Ag–Cu lead-free soldered joints. J Alloys Compd. 2003;352:226–36.
15. Zou HF, Zhang ZF. Ductile-to-brittle transition induced by increasing strain rate in Sn–3Cu/Cu joints. J Mater Res. 2008;23:1614–7.
16. Hu GJ, Goh KY, Judy L. Micromechanical analysis of copper trace in printed circuit boards. Microelectron Reliab. 2011;51:416–424.
17. Sinkovics B, Krammer O. Board level investigation of BGA solder joint deformation strength. Microelectron Reliab. 2009;49:573–8.

18. Zhang QK, Tan J, Zhang ZF. Fracture behaviors and strength of Cu_6Sn_5 intermetallic compounds by indentation testing. J Appl Phys. 2011;110:014502.
19. Zhang QK, Zhang ZF. Influences of reflow time and strain rate on interfacial fracture behaviors of Sn–4Ag/Cu solder joints. J Appl Phys. 2012;112:064508.
20. Zhang QK, Zhang ZF. In-situ observations on fracture behaviors of Cu–Sn IMC layers induced by deformation of Cu substrates. Mater Sci Eng A. 2011;530:452–61.
21. Yoon JW, Kim SW, Jung SB. Interfacial reaction and mechanical properties of eutectic Sn–0.7Cu/Ni BGA solder joints during isothermal long-term aging. J Alloys Compd. 2005;391:82–9.
22. Kim SW, Yoon JW, Jung SB. Interfacial reactions and shear strengths between Sn–Ag–based Pb-Free Solder Balls and Au/EN/Cu Metallization. J Electron Mater. 2004;33:1182–9.
23. Kerr M, Chawla N. Creep deformation behavior of Sn–3.5Ag solder/Cu couple at small length scales. Acta Mater. 2004;52:4527–35.
24. Shang PJ, Liu ZQ, Pang XY, Li DX, Shang JK. Growth mechanisms of Cu_3Sn on polycrystalline and single crystalline Cu substrates. Acta Mater. 2009;57:4697–706.
25. Zou HF, Yang HJ, Zhang ZF. Morphologies, orientation relationships and evolution of Cu_6Sn_5 grains formed between molten Sn and Cu single crystals. Acta Mater. 2008;56:2649–62.
26. Zhang QK, Zhang ZF. Fracture mechanism and strength-influencing factors of Cu/Sn–4Ag solder joints aged for different times. J Alloy Compd. 2009;485:853–61.
27. Lang F, Tanaka H, Munegata O, Taguchi T, Narita T. The effect of strain rate and temperature on the tensile properties of Sn–3.5Ag solder. Mater Charact. 2005;54:223–9.
28. Bai N, Chen X, Fang Z. Effect of strain rate and temperature on the tensile properties of tin-based lead-free solder alloys. J Electron Mater. 2008;37:1012–9.
29. Date M, Shoji T, Fujiyoshi M, Sato K, Tu KN. Ductile-to-brittle transition in Sn–Zn solder joints measured by impact test. Script Mater. 2004;51:641–5.
30. Zhu QS, Zhang ZF, Shang JK, Wang ZG. Fatigue damage mechanisms of copper single crystal/Sn–Ag–Cu interfaces. Mater Sci Eng A. 2006;435–436:588–94.
31. Yao P, Liu P, Liu J. Effects of multiple reflows on intermetallic morphology and shear strength of SnAgCu–xNi composite solder joints on electrolytic Ni/Au metallized substrate. J Alloys Compd. 2008;462:73–9.
32. Zhang ZF, Wang ZG. Grain boundary effects on cyclic deformation and fatigue damage. Prog Mater Sci. 2008;53:1025–99.
33. Kumar KS, Van Swygenhoven H, Suresh S. Mechanical behavior of nanocrystalline metals and alloys. Acta Mater. 2003;51:5743–74.
34. Zhang ZF, Wang ZG. Dependence of intergranular fatigue cracking on the interactions of persistent slip bands with grain boundaries. Acta Mater. 2003;51:347–64.
35. Meyers MA, Mishra A, Benson DJ. Mechanical properties of nanocrystalline materials. Prog Mater Sci. 2006;51:427–556.
36. Head AK. Dislocation group-dynamics. 6. Release of a pile-up. Philos Mag. 1973;27:531–539.
37. Larsson AK, Stenberg L, Lidin S. The superstructure of domain-twinned eta'—Cu_6Sn_5. Acta Crystall B. 1994;50:636–43.
38. Kao C. Microstructures developed in solid-liquid reactions: using Cu–Sn reaction, Ni–Bi reaction, and Cu–In reaction as examples. Mater Sci Eng A. 1997;238:196–201.
39. Cottrell AH. Dislocations and plastic flow in crystals. London: Oxford University Press; 1953.
40. Armstrong R, Codd I, Douthwaite RM, Petch NJ. The plastic deformation of polycrystalline aggregates. Philos Mag. 1962;7:45–58.
41. Zeng K, Tu KN. Six cases of reliability study of Pb-free solder joints in electronic packaging technology. Mater Sci Eng R. 2002;38:55–105.

Chapter 3
Tensile-Compress Fatigue Behavior of Solder Joints

3.1 Introduction

Since soldering in the microelectronic assembly not only provides the electronic connection, but also ensures the mechanical reliability of solder joints under the complex service conditions, it has been recognized that one of the major concerns for the integrity of the solder interconnection is the mechanical property of the solder/substance interfaces [1]. In particular, with the trend that the solder joints are expected to service in more dynamic environments where the stress and strain distributions change with time, not only the tensile or shear properties but also the fatigue resistance of the solder joints become the great significance for electronic reliability [1, 2]. Furthermore, the continuing miniaturization of solder joints has set up higher standards for their mechanical reliability.

In recent years, there have been many investigations concerning the fatigue properties of the lead-free solders and solder joints [3–9]. However, the correlative data of that available in the literature are still limited and always from different resources. There is a complex evolution of the solder/substrate interface when the solder joints are in service, namely, the growth of interfacial IMC layer and coarsening of the solder's microstructure [10], while the influence of interfacial evolution on the fatigue resistance and failure mechanism of solder joints are lack of investigation.

In the present study, therefore, the authors have concentrated their efforts to reveal the fatigue behaviors of a series of the Cu/Pb-free solder joints. Both the as-soldered and thermal-aged interfaces were tested, in order to get a comprehensive understanding on the interfacial fatigue failure mechanisms. Some reliable data of fatigue lives were tested to make a quantitative comparison. The fracture behaviors of as-soldered and thermal-aged solder joints were compared to show the influence of thermal aging on that. The evolutions of microstructure and mechanical properties of the Sn–Ag and Sn–Bi solder alloys during the aging process were also observed to make a full-scale discussion. Based on the experimental results, the fatigue failure mechanisms and factors influencing the interfacial fatigue were discussed.

© Springer-Verlag Berlin Heidelberg 2016
Q. Zhang, *Investigations on Microstructure and Mechanical Properties of the Cu/Pb-free Solder Joint Interfaces*, Springer Theses, DOI 10.1007/978-3-662-48823-2_3

3.2 Experimental Procedure

The Cu substrate material used in this study was prepared from oxygen-free high-conductivity (OFHC) Cu of 99.999 % purity by the Bridgman method in a horizontal furnace. The solder alloys chosen were Sn–4Ag (wt%), Sn–58Bi (wt%), and Sn–37Pb; the Cu/Sn–37Pb solder joints were investigated as reference. All the solder alloys were prepared by melting high-purity (>99.99 %) tin and Ag, Bi, Pb, and Cu at 800 °C for 2 h in vacuum, and the smelting process was operated in a vacuum furnace. To reveal the influence of thermal aging on tensile properties of the solders, the air-cooled Sn–58Bi and Sn–4Ag bulk solder alloys were cut into standard tensile samples (see Fig. 3.1a), aged at 120 and 180 °C for different times and the side surfaces were ground. The tensile tests were performed with an Instron E1000 fatigue testing machine under a strain rate of 1.25×10^{-4} s^{-1} at 20 °C in air. Three samples were tested at the same condition to get an average strength.

The solder joints for mechanical property tests were prepared by reflow soldering of two Cu plates. The surfaces of the Cu plates for soldering were first polished with a diamond polishing agent, a soldering flux was then dispersed on the polished area, and finally a solder alloy sheet was placed on the paste to ensure sufficient wetting reaction. The prepared samples were kept in an oven at a constant temperature for a few minutes and then cooled down in air. Some of the soldered samples were isothermally aged for different times. The reflowing, aging temperatures and times are listed in Table 3.1. After cooling, both the as-reflowed and aged samples were spark cut into fatigue specimens, and the side surfaces were ground and carefully polished for microstructure observations of the Cu/solder interface. The dimensions of the finished fatigue specimens are presented in Fig. 3.1b.

The stress-controlled fatigue tests were carried out by Instron 8871 fatigue testing machines under a symmetrical sinusoid waveform ($R = -1$) with a frequency

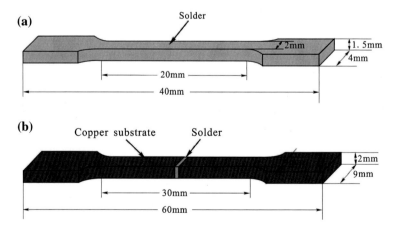

Fig. 3.1 Configuration of **a** tensile specimens of bulk solder alloy and **b** tensile and fatigue specimens. Reprinted from Ref. [11], Copyright 2010, with permission from Elsevier

Table 3.1 Reflowing, aging temperatures and times of different solder joints

Solder joints	Reflowing temperature (°C)	Reflowing time (min)	Aging temperature (°C)	Aging time (day)
Sn–4Ag/Cu	260	5	180	4, 16
Sn–37Pb/Cu	220	5	160	7
Sn3.8Ag0.7Cu/Cu	240	5	170	7
Sn–58Bi/Cu	200	3	120	4, 7, 9

of 2 Hz. The fatigue lives of samples were tested at different stress amplitudes and three samples were tested under each condition. The stress amplitude was increased until the fatigue life decreased to about 10^3 cycles. For some samples, cyclic deformation was interrupted at different cycles, and then the interfacial deformation and cracking behaviors were observed with a ZEISS Supra 35 SEM to reveal the interfacial deformation behavior and fatigue fracture mechanisms.

3.3 Experiment Results

3.3.1 Tensile and Fatigue Behavior of Sn–4Ag/Cu Interface

The interfacial microstructure and morphologies of IMC grains in the solder joint reflowed for 3 min at 260 °C and then aged at 180 °C for different times are shown in Fig. 3.2. During the initial soldering process, the Cu_6Sn_5 IMC displayed a scallop-like morphology forming along the interface of the Sn–4Ag/Cu joints, and the average thickness is about 2 μm, as shown in Fig. 3.2a. Figure 3.2b shows the

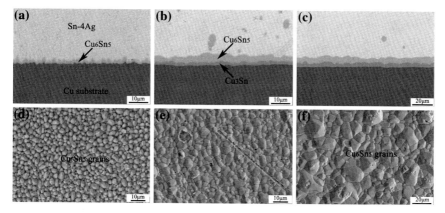

Fig. 3.2 Morphologies of interfacial microstructure and interfacial IMCs: after aging for different times: **a, b** as-soldered sample; samples aged at 180 °C for **c, d** 4 days and **e, f** 16 days. Reprinted from Ref. [12], with kind permission from Springer Science+Business Media

morphology of the Cu_6Sn_5 grains in the as-soldered sample. It can be seen that the Cu_6Sn_5 grains are spherical. After aged at 180 °C for 4 days, the IMC/solder interface became quite flat, and the IMC thickness increased to about 6 μm, as shown in Fig. 3.2c. Meanwhile, the scallop-type Cu_6Sn_5 layer was changed to planar type with duplex layers of Cu_6Sn_5/Cu_3Sn and the spherical Cu_6Sn_5 grains became typical equiaxed ones (see Fig. 3.2d). The flattening process requires that the growth of the IMC should be faster in the valleys of the scallop than in the peaks [12]. After aged for 16 days, the thickness of the interfacial IMC increased to about 11 μm and there was no obvious coarsening of the IMC/solder interface, as shown in Fig. 3.2e. The morphology of Cu_6Sn_5 grains is similar to that of the samples aged for 4 days, and only the grain size becomes larger, as in Fig. 3.2f. The growth of the interfacial IMC layers during aging process is controlled by a diffusion mechanism, and the growth rate of the Sn–4Ag/Cu interface is approximate to the SnAgCu/polycrystalline Cu and the SnAgCu/Cu single crystal interfaces [13–16]. In addition, comparing Fig. 3.2b with Fig. 3.2a, it is found that there is a little coarsening of the needle-like Ag_3Sn grains in the solder, which may induce a slight decrease in yield strength of the solder.

Figure 3.3 shows the tensile stress–displacement curves of the solder joints aged for different times. It shows that the tensile strength (σ_b) of the solder joints was dropped from 59.2 MPa (as-soldered samples) to 46.8 MPa (samples aged for 4 days) and 38.0 MPa (samples aged for 16 days), respectively. This tendency is in agreement with many previous results [17, 18]. In addition, the yield strengths of those samples are in the range of 15–20 MPa and the samples exhibited strong strain hardening during the tensile test. Since the solder is only a small fraction of the whole sample, a tensile stress–displacement curve of the whole sample should be quite similar to that of the Cu single crystal; thus the yield and strong strain hardening behavior should mainly result from plastic deformation of the Cu single crystal. Although the stress amplitudes are always higher than the yield strength of

Fig. 3.3 Tensile
stress–displacement curves
of as-soldered and aged
Sn–4Ag/Cu solder joints.
Reprinted from Ref. [12],
with kind permission from
Springer Science+Business
Media

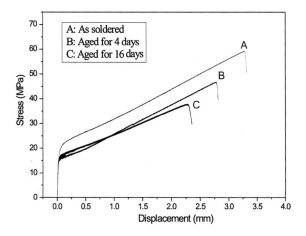

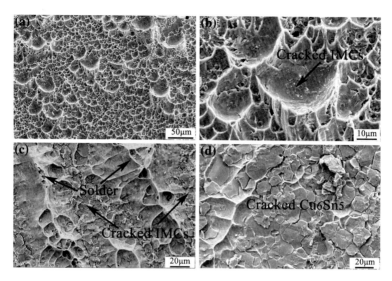

Fig. 3.4 Tensile fracture morphologies of Sn–4Ag/Cu solder joints: **a** microscopic and **b** magnified dimple fracture morphology of the as-soldered sample; **c** mixed fracture morphology of the sample aged for 4 days; **d** cleavage fracture morphology of the sample aged for 16 days. Reprinted from Ref. [12], with kind permission from Springer Science+Business Media

the Cu single crystal during a fatigue test ($\geq$20 MPa), the Cu substrate should not display much plasticity due to the strong work hardening.

Figure 3.4 shows the tensile fracture morphologies of the samples aged for different times, from which one can conjecture the microscopic tensile fracture mechanism. The as-soldered sample exhibited a typical ductile fracture feature as indicated by the numerous dimples in Fig. 3.4a. There are some cracked IMCs at the bottom of the dimples (see Fig. 3.4b), indicating that the micro-cracks first initiate at the top of the IMC layer and then propagate into the solder along a special angle to form a dimple. In contrast, the samples aged for 16 days displayed a typical transgranular brittle fracture in the interior of the Cu_6Sn_5 grains (see Fig. 3.4d). The fracture mechanism of the samples aged for 4 days is complicated. According to the fractographies shown in Fig. 3.4c, the samples broke into a mixed mode or a ductile-to-brittle transition; this occurred because both ductile dimples and brittle IMCs coexisted on the fracture surface. As a result, it can be predicated that the fracture mode may transform from ductile to brittle when the interfacial IMC thickness increases to about 6 μm, which is well consistent with the decrease in the tensile strength.

As the Young's modulus, yield strength, and hardness of Cu_6Sn_5 and Cu_3Sn phases are much higher than those of the solder alloy and Cu substrate, the external stress cannot be the only reason for the fracture of the interfacial IMC during tensile test. In fact, due to the decrease in the volume during continuous growth of Cu_6Sn_5 and Cu_3Sn layers, there is residual stress accumulated at the interface of solder/Cu [19–21]. Besides, the simulation of stress distribution indicates that there is severe

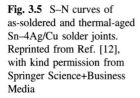

Fig. 3.5 S–N curves of as-soldered and thermal-aged Sn–4Ag/Cu solder joints. Reprinted from Ref. [12], with kind permission from Springer Science+Business Media

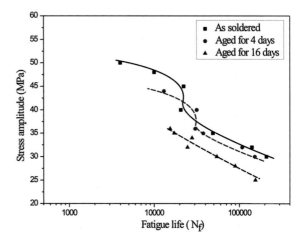

stress concentration at the corner of the Cu/Cu_6Sn_5 interface, which was induced by the strain incompatibility [22]. Therefore, for the aged samples, the interfacial IMCs are easy to fracture at a low stress and exhibit a brittle fracture feature.

The stress amplitude (σ_a)–fatigue life (N_f) relationship (S–N curve) of the Sn–4Ag/Cu solder joints aged for different times is shown in Fig. 3.5. When the stress amplitude is low (e.g., $\sigma_a \leq 35$ MPa), the fatigue life displays an approximately exponential increase with decreasing stress amplitude. Besides, it is obvious that fatigue life of the as-soldered samples is close to that of the samples aged for 4 days, but is obviously longer than that of the samples aged for 16 days. This indicates that a short-time aging (e.g., 4 days) might slightly affect the fatigue life of the joint samples, but a long-time aging (e.g., 16 days) may markedly deteriorate the fatigue properties of the joint samples. The fatigue failure processes and effect of adhesive strength on fatigue life will be discussed later in detail.

When the stress amplitude is higher (e.g., $\sigma_a \geq 40$ MPa), it is obvious that the fatigue life abnormally deviates from the conventional low-cycle fatigue life range. When the stress amplitude is between 40 and 45 MPa, the S–N curve has a nearly upright ascent, and then descends with increasing stress amplitude, which is similar to that of a Pb-rich solder alloy [23]. The abnormal range of the as-reflowed samples is a little bit higher than that of the samples aged for 4 days. The abnormal fatigue life increase under higher stress amplitude may be caused by yield of the solder, because the yield strength of Sn–4Ag solder is about 40–50 MPa under the current strain rate [1, 24]. When the stress amplitude is higher than the yield strength, the solder will yield to release the strain incompatibility between the solder and the interfacial IMC. It has been reported that aging often induces a little drop in the yield strength due to the coarsening of the solder alloy [25, 26]. Therefore, the S–N curve of samples aged for 4 days deviates in the lower stress amplitude range compared to that of the as-soldered ones. For the samples aged for 16 days, this phenomenon does not exist, because the stress amplitude is lower than the yield strength of the solder. When the stress amplitude increases to a value close to the tensile strength, the fatigue life of the samples will decrease rapidly to less than 1000, which may correspond to an obvious transition of the fatigue fracture mechanisms.

To reveal the interfacial fatigue crack initiation and propagation behavior under tensile-compress loadings, some samples were fatigued for fixed cycles for side surface observations. The results showed that the samples aged for different times have similar crack initiation mechanism. In Fig. 3.6, an as-soldered sample tested at stress amplitude of 35 MPa was employed to reveal the initiation mechanism. Figure 3.6a shows the side surface morphology of the as-soldered sample after 5000 cycles. Close to the interfacial IMC, severe plastic deformation of the solder was observed due to the stress concentration at the solder/IMC interface. In fact, such stress concentration has been proved by the simulation of stress distribution at the Cu/solder interface [22]. Considering the great difference in mechanical properties between the solder and the IMC, especially the hardness, it is easy to understand the severe strain incompatibility and stress concentration that exists near the interfaces [19, 20]. After 10,000 cycles, some superficial micro-cracks appeared at the solder/Cu_6Sn_5 interface, as in Fig. 3.6b. With continuous increase of cyclic number, plastic deformation at the solder/IMC interface became more and more severe. The micro-cracks linked up with each other to form a long initial crack, and propagated into the solder near the solder/Cu_6Sn_5 interface, as shown in Fig. 3.6c. With further increasing cycles, the fatigue cracks kept propagating (Fig. 3.6d). In addition, some reports indicated that the plate-like Ag_3Sn grains can be regarded as a strengthening phase to improve the resistance to fatigue cracking [27]. In this study, inhomogeneous plastic deformation with a core of plate-like Ag_3Sn grains in the solder

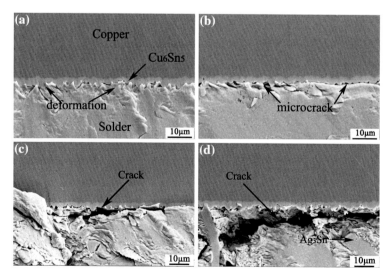

Fig. 3.6 Fatigue deformation and damage morphologies at the as-soldered Sn–4Ag/Cu joint interface cycled at 35 MPa for **a** 5000, **b** 10,000, **c** 20,000, and **d** 30,000 cycles. Reprinted from Ref. [12], with kind permission from Springer Science+Business Media

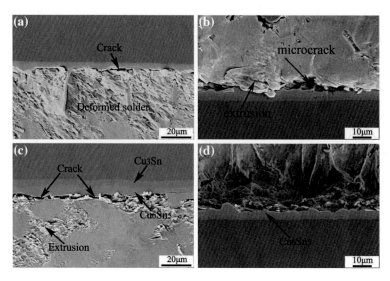

Fig. 3.7 Fatigue damage morphologies at the aged Sn–4Ag/Cu joint interface: **a**, **b** aged for 4 days and cycled at 35 MPa for 10,000 cycles; **c** aged for 16 days and cycled at 30 MPa for 10,000 cycles; **d** side surface of the fractured sample aged for 4 days. Reprinted from Ref. [12], with kind permission from Springer Science+Business Media

adjacent to the interface was observed, as shown in Fig. 3.6d, which may decrease the fatigue life of the solder joints by promoting the crack initiation. However, as there are only a few plate-like Ag_3Sn grains at the joint interface, the plate-like Ag_3Sn compound may actually have little effect on fatigue crack initiation.

The fatigue deformation and damage behavior of the aged interface are shown in Fig. 3.7. Figure 3.7a, b shows the interfacial morphology of the solder joint aged for 4 days and then deformed at 35 MPa for 10,000 cycles. It can be found that the interfacial fatigue deformation and crack initiation mechanism change little compared with the as-soldered interface; there is still obvious strain concentration around the interface, and the micro-cracks appear at the solder/Cu_6Sn_5 interface. In Fig. 3.7b, both the extrusion of solder and cracking along the solder/Cu_6Sn_5 interface can be observed. After aged for 16 days, the fatigue crack still initiates at the solder/Cu_6Sn_5 interface, but the prominent Cu_6Sn_5 grain may fracture (see Fig. 3.7c). In general, the plastic deformation of the solder close to the joint interface is severe, the initial fatigue crack of the Sn–4Ag/Cu solder joint appears at the solder/Cu_6Sn_5 interface rather than in the solder, and thus there is no residual solder at the side surface of the fracture sample, as in Fig. 3.7d.

Fatigue crack propagation path and fracture mechanism can be predicated from the fracture morphologies of solder joints. Observations showed that the samples aged for different times and tested at different stress amplitudes have similar fatigue fracture surfaces. Figure 3.8 shows the typical fracture morphologies of the fatigued samples. The macroscopic image of fatigue fracture surface is illustrated in Fig. 3.8a. It is obvious that there are two regions in general, i.e., (1) fatigue crack

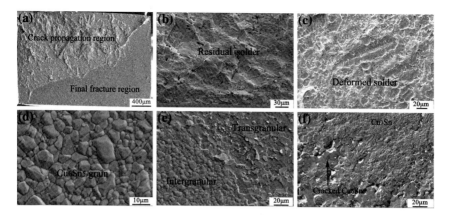

Fig. 3.8 Fatigue fracture surfaces of Sn–4Ag/Cu solder joints: **a** macroscopic image; **b** fracture morphologies, showing fatigue crack source; **c** fatigue crack propagation region; **d** intergranular fatigue fracture along the Cu_6Sn_5/solder interface; **e** transition region from intergranular to transgranular fatigue fracture; **f** transgranular fracture along the Cu_6Sn_5/Cu_3Sn interface. Reprinted from Ref. [12], with kind permission from Springer Science+Business Media

propagation region covered by deformed solder and (2) final fracture region with flat morphology on a macroscale. As the fatigue cracks often initiated at the sample edge, the fracture surface in this region is quite smooth due to the effects of friction and crack closure, as in Fig. 3.8b. In contrast, the fatigue crack propagation region is quite rough and covered by a layer of deformed solder, as shown in Fig. 3.8c. Combined with the side surface observations in Fig. 3.7, it is obvious that fatigue crack propagated in the solders close to the solder/Cu_6Sn_5 interface. Little IMC was found in the former two regions. Another difference from the fatigue fracture surface of common engineering materials is that there is no fatigue striation observed in the fatigue crack propagation region, as in Fig. 3.8c. The morphology of the final fracture region is a little bit complex. Although all the samples finally failed in a brittle mode, there are always two types of fracture modes observed in the same sample, i.e., intergranular fracture and transgranular fracture. Figure 3.8e shows the interface of the intergranular fracture and transgranular fracture. For the intergranular fracture, cracking often occurred at the interface of Cu_6Sn_5 grains and solder; and the interfacial Cu_6Sn_5 grains remained intact after the final fracture. In contrast, for the transgranular fracture mode, cracking occurred in the interior of IMCs. For the samples aged for 16 days or subjected to higher stress amplitudes, some Cu_3Sn grains can be found at the final fracture surface because it is too brittle (see Fig. 3.8f).

3.3.2   Tensile and Fatigue Behavior of Sn–58Bi/Cu Interface

The interfacial morphologies of the Sn–Bi/Cu joints are shown in Fig. 3.9. For the as-soldered joint exhibited in Fig. 3.9a, an Cu_6Sn_5 layer of about 1 μm thick was observed between the Cu substrate and Sn–Bi solder. After aged at 120 °C for 5 days, the IMC layer became evidently thick, and a very thin Cu_3Sn layer was observed between the Cu substrate and the Cu_6Sn_5 layer, as shown in its backscattered electron image (see Fig. 3.9b). Meanwhile, noticeable coarsening of the microstructure of the Sn–Bi eutectic was observed, which may induce a change in the mechanical properties of the solder alloy. The interfacial morphologies of the samples aged for 10 and 13 days are shown in Fig. 3.9c, d, respectively. It is obvious that the IMC thickness increases with increasing aging time, while the alteration of the interfacial morphology was little. Similar to the reports before [28–30], discontinuous Bi particles (the bright spots) were also observed at these long-term aged IMC/Cu interfaces, which mainly came from the Bi diffusion of the Sn–Bi solder through the IMC layers during soldering and aging processes [31]. Because the Cu–Bi system is completely immiscible [32], the diffusion of Bi atoms into the pure Cu substrate should be extremely difficult and the Bi atoms can easily segregate at the Cu_3Sn/Cu interface to form the particles. In addition, compared to the microstructure of the Sn–Bi solder in Fig. 3.9b–d, it was found that the solder shows little coarsening during the later stage of aging, indicating that coarsening of the solder alloy occurred only at the early stage of aging.

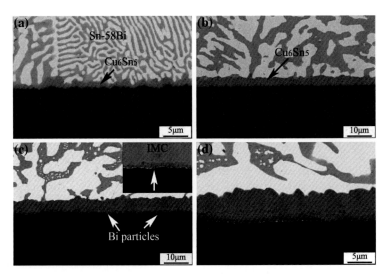

Fig. 3.9 Interfacial morphologies of Sn–58Bi/Cu solder joints (backscattered electron images): **a** as-soldered and aged for **b** 5 days, **c** 10 days, **d** 13 days. Reprinted from Ref. [33], Copyright 2010, Materials Research Society

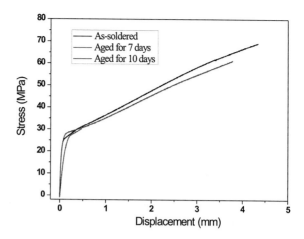

Fig. 3.10 Tensile stress–strain curves of as-soldered and aged Sn–58Bi/Cu solder joints

The tensile stress–strain curves of the Sn–Bi/Cu solder joints reflowed for different times are shown in Fig. 3.10. Since the substrate material is the same, the tensile curves of the Sn–Bi/Cu and Sn–Ag/Cu solder joints are similar. As in the curve, the yield strength of the Sn–Bi/Cu solder joint is about 20–30 MPa, and there is a long strain hardening stage, which is actually the yield and strain hardening of the Cu substrate. Since the Sn–Bi solder is brittle, the Sn–Bi/Cu solder joints fractured when the stress increased to a certain value. As the Cu substrate is still in its hardening stage when the solder joint fractures, it can be predicated that under tensile-compress fatigue loadings, the Cu substrate will harden rapidly rather than fracture. Besides, the strength of the solder joint reflowed for 10 days decreases to about 30 MPa, showing the "aging brittlement."

The tensile strengths of the Sn–Bi/Cu solder joints aged for different times are shown in Fig. 3.11. The tensile strength of the as-soldered solder joint is about 70 MPa and decreases to 60 MPa after aged for 7 days; the decrease amplitude is

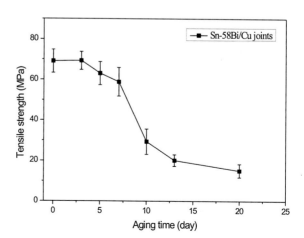

Fig. 3.11 Tensile strength of as-soldered and aged Sn–58Bi/Cu solder joints. Reprinted from Ref. [33], Copyright 2010, Materials Research Society

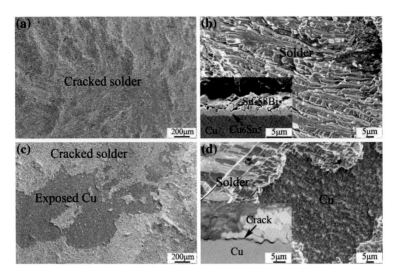

Fig. 3.12 Tensile fracture surfaces of the Sn–58Bi/Cu solder joints: **a** Macroscopic and **b** microscopic morphologies of as-soldered samples; **c** Macroscopic and **d** microscopic morphologies of solder joints aged for 10 days

low, which should be induced by the coarsened microstructure of the solder. However, the tensile strength aged for 10 days decreases to 30 MPa, and further decreases to 20 MPa. In the tensile strength–aging time curve, the tensile strength decreases sharply when the aging time is 7–10 days, indicating that there should be an essential transform in fracture mechanism. It is generally recognized that the brittlement is related to the Bi segregation at the Cu_3Sn/Cu interface [28, 30].

The tensile fracture surface and side surface of the as-soldered and aged Sn–Bi/Cu joint are shown in Fig. 3.12. Figure 3.12a, b shows the fracture morphologies of the as-soldered solder joint. In macroscale the fracture surface is coarsened, and the fractured solder can be observed. The side surface also indicates that fracture occurs inside the solder close to the joint interface. After aged for 10 days, the strength of the Sn–Bi/Cu solder joint decreases obviously, and the macroscopic fracture surface is composed of two regions with obvious difference in appearance. The bright area is covered by solder, and the dark area is naked Cu substrate, indicating that there is a change in fracture mechanism; the fracture location transforms from the solder close to the interface to the Cu_3Sn/Cu interface, as shown in Fig. 3.12c, d. It has been reported that appearance of the Bi atoms at the interface leads to the decrease in the interfacial adhesive strength, which further decreases the strength of the Sn–Bi/Cu interface [34].

For both the as-soldered and aged Sn–Bi/Cu solder joints, the fatigue life also displays an approximately exponential increase with decreasing stress amplitude, as in Fig. 3.13. It has been reported that the Sn–Bi/Cu solder joints have low fatigue resistance, because of the inferior ductility of the Sn–Bi solder [35], whereas the results in this study show that they have comparable fatigue resistance than the

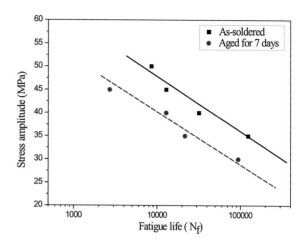

Fig. 3.13 S–N relationship of as-soldered and thermal-aged Sn–58Bi/Cu solder joints

Sn–Ag/Cu solder joints at the same stress amplitude. Although the solders have poor ductility, the higher tensile strength of the Sn–Bi/Cu solder joints conceals shortage to some extent at low stress amplitude, making their fatigue resistance comparable with the Sn–Ag/Cu solder joints. However, as the Sn–Bi/Cu solder joints have much higher fracture strength than the Sn–Ag/Cu joints, they do have poor fatigue resistance at the same stress ratio (σ_a/σ_f). The Sn–Bi/Cu solder joints did not exhibit an upright ascent phenomenon because the applied stress amplitude is far below the yield strength of the solder and the curves may never reach the turning point. In addition, the decrease in fatigue life of the Sn–Bi/Cu solder joints after aging is much obvious. After aging for over 9 days, the Sn–Bi/Cu solder joints exhibited a totally brittle feature.

Figure 3.14 shows the deformation behavior of the Sn–58Bi/Cu solder joints, and also one sample is chosen as example for each aging condition. The interfacial morphology of the as-soldered sample deformed at 50 MPa for 5000 cycles is shown in Fig. 3.14a, in which long fatigue crack was observed along the solder/IMC interface. Due to the high yield strength of the Sn–Bi alloy, deformation of the solder close to the solder/IMC interface is quite slight. Figure 3.14b shows the interfacial morphology of the sample aged for 7 days and deformed at 35 MPa for 2×10^4 cycles. Although the coarsening of the solder was obvious and the interfacial IMC layer was much thicker, fatigue crack still initiated exactly along the solder/IMC interface. However, the solder joints aged for over 9 days fractured along the IMC/Cu interface at very low stress amplitude, as in Fig. 3.14c, accompanying with a sharp decrease in fatigue life. The essential transition in fracture mechanism of the long-term aged Sn–Bi/Cu solder joints is induced by the interfacial Bi segregation.

The fatigue fracture surface before the embrittlement occurs is shown in Fig. 3.15. Figure 3.15a shows the typical fatigue fracture surface of the Sn–Bi/Cu solder joints prior the brittle fracture, which is also composed of two regions: the first region is the naked Cu_6Sn_5 grains, with little residual solder; the second area is

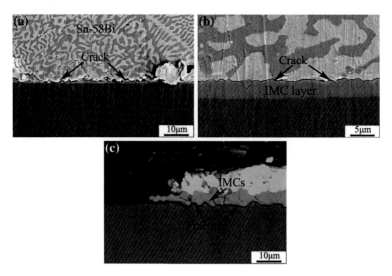

Fig. 3.14 Deformation and crack initiation behaviors of the Sn–58Bi/Cu joint interface aged for different times: **a** as-soldered sample, deformed at 50 MPa for 5000 cycles; **b** sample aged for 7 days, deformed at 35 MPa for 20,000 cycles; **c** side surface of the fractured sample aged for 9 days. Reprinted from Ref. [11], Copyright 2010, with permission from Elsevier

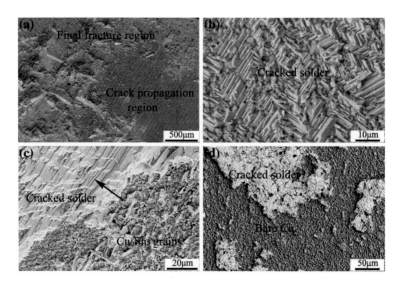

Fig. 3.15 Fatigue fracture surfaces of Sn–58Bi/Cu solder joints: **a** macroscopic image; **b** morphology of final fracture surface; **c** transition area from crack propagation region to final fracture region; **d** fracture surface of sample aged for 10 days. Reprinted from Ref. [11], Copyright 2010, with permission from Elsevier

the same to the tensile fracture surface, with a thin layer of solder on it (see Fig. 3.15b). Based on this, it is predicated that the region covered by solder is the final fracture zone, and the region covered by Cu_6Sn_5 grains is the crack propagation zone. The first region is covered by some perfect Cu_6Sn_5 grains, with little residual solders, as in Fig. 3.15c. In contrast, the microscopic morphology of the second region is covered by a thin layer of cracked solder, the same as the tensile fracture surfaces of the solder joints [33], and thus it is considered to be the final fracture region and the former one is the crack propagation region. It can be therefore concluded that for the Sn–Bi/Cu solder joints, the fatigue cracks initiate and propagate along the solder/Cu_6Sn_5 interface and finally fracture inside the solder close to the interface. The fracture surface of the long-term aged sample is mainly the exposed Cu, with little cracked solder remnants (see Fig. 3.15d), which is consistent well with the side surface observation results.

3.3.3 Fatigue Behavior of Sn–37Pb/Cu Interface

The interfacial morphologies of the Sn–37Pb/Cu interface are shown in Fig. 3.16. The as-soldered interface is shown in Fig. 3.16a, in which the IMC thickness is about 1 μm. Compared with the as-soldered Sn–4Ag/Cu interface, the IMC layer at the Sn–37Pb/Cu interface is obviously thinner, which should be induced by the low soldering temperature and the lower Sn content. After aging for 7 days, the IMC thickness increases to 6 μm, as in Fig. 3.16b, and the thicknesses of Cu_6Sn_5 and Cu_3Sn are about 3.5 and 2.5 μm, respectively. It should be noticed that a discontinuous Pb-rich layer appears around the IMC close to the solder, because the Sn and Cu atoms in this region are consumed, making the Pb content to increase.

The S–N curves of the Sn–Pb/Cu solder joints are very similar to that of the Sn–Ag/Cu joints. At low stress amplitude, the life cycles exponentially increase with decrease in stress amplitude, as shown in Fig. 3.17. When the stress amplitude is about 40 MPa, there is also an abnormal increase in fatigue life; the reason should

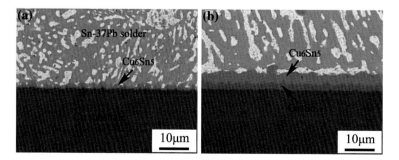

Fig. 3.16 Morphologies of the Sn–37Pb/Cu joint interface: **a** as-soldered sample, **b** samples aged at 160 °C for 7 days

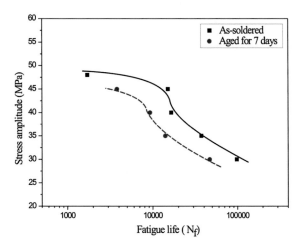

Fig. 3.17 S–N curves of as-soldered and thermal-aged Sn–37Pb/Cu solder joints. Reprinted from Ref. [11], Copyright 2010, with permission from Elsevier

be similar to that of the Sn–4Ag/Cu solder joints. Since the yield strength of the Sn–37Pb solder is low, the abnormal region is also lower. Besides, the Sn–37Pb/Cu solder joints show the phenomenon during the thermal fatigue process [23]. In general, the fatigue life of the Sn–37Pb/Cu solder joint is obviously lower than the Sn–4Ag/Cu solder joint, and also little lower than the Sn–58Bi/Cu solder joints, which affirm that the two lead-free solders have higher fatigue resistance than the Sn–37Pb solder.

The fatigue damage behavior of the Sn–37Pb/Cu solder joints is shown in Fig. 3.18. Figure 3.18a shows the interfacial morphology of an as-soldered sample deformed at 45 MPa for 7000 cycles, and it can be found that crack is initiated inside the solder close to the solder/IMC interface. The edge of the fracture surface of the as-soldered sample is covered by deformed solder, as in Fig. 3.18b, indicating that cracks propagate into the solder near the solder/IMC interface once it forms. Similar to the Sn–Ag/Cu solder joints, the crack initiation should be induced by the strain localization, but easier as the Sn–37Pb solder has low yield strength and superior ductility. After aging for 7 days, there is little change in fatigue failure mechanism. According to Fig. 3.18c, d, crack still initiates and propagates inside the solder, but more close to the solder/IMC interface, because IMC grains were observed at the edge of the fracture surface. It is therefore concluded that the crack initiation behavior of the Sn–Pb/Cu solder joints is similar to that of the Sn–Ag/Cu joints in general. However, the fatigue fracture surfaces are a bit different.

The fatigue fracture surface of the Sn–Pb/Cu solder joints is shown in Fig. 3.19. As shown in Fig. 3.19a, the macroscopic view of the fracture surface of the Sn–Pb/Cu interface is rough and fracture step is observed, indicating that the fatigue cracks propagate into the solder along a special angle during the propagate process. The final fracture is a step fracture. The crack propagation region extends from the edge of the sample to the fracture stage. In microscopic, the crack propagation region is covered by deformed solder, as in Fig. 3.19b. Figure 3.19c shows the microscopic view of the interface of the two regions; it is obvious that the two

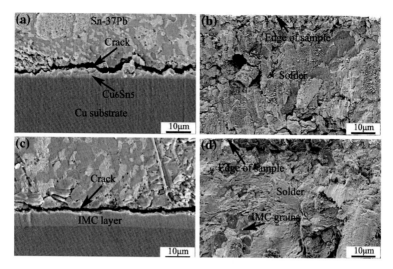

Fig. 3.18 Fatigue crack initiation behaviors of Sn–37Pb/Cu interfaces: **a** as-soldered sample, $\sigma_a = 45$ MPa, $N = 7000$; **b** fracture morphology at the edge of as-soldered sample; **c** sample aged for 7 days, $\sigma_a = 35$ MPa, $N = 10^4$; **d** fracture morphology at the edge of the aged sample. Reprinted from Ref. [11], Copyright 2010, with permission from Elsevier

Fig. 3.19 Fatigue fracture morphology of Sn–37Pb/Cu solder joints: **a** macroscopic fracture surface, and **b** transition region of crack propagation region and final fracture region, microscopic fracture surfaces of **c** crack propagation region and **d** final fracture region. Reprinted from Ref. [11], Copyright 2010, with permission from Elsevier

regions show different morphologies in macro. Dimples were observed at the final fracture region, as in Fig. 3.19d. Therefore, it is predicated that the fatigue cracks propagate inside the solder and the final fracture occurs inside the solder in a dimple mode due to its superior ductility.

3.3.4 Evolution in Strength of Solder During Aging

As the solder coarsened during the aging process, it is necessary to show its influence on the mechanical properties of the solders. Evolutions in tensile strengths of the Sn–4Ag and Sn–58Bi solder alloys during the aging process are shown in Fig. 3.20. The Sn–58Bi solder has much higher yield strength than the Sn–4Ag solder at the current strain rate. Moreover, though there is obvious coarsening in the solders, the tensile strength only decreases slightly and becomes steady at the later aging process, which is consistent with some earlier reports [25]. Since the melting points of the Sn-based solders are very low, its homogeneous temperature is quite high at 120 and 160 °C; thus the coarsened process is completed in a very short time.

The fatigue fracture processes of different Cu/solder interfaces are summarized in Fig. 3.21. For a typical fatigue fracture, the fatigue fracture processes consist of the crack initiation and propagation stages. According to the foreside discussion, for all the solder joints, micro-cracks initiate around the solder/IMC interface due to the strain localization. The deformation mismatch between solder and the interfacial IMC layer is the essential reason of fatigue crack initiation. However, the propagation paths are a little bit different, affected by the mechanical properties of the solder, as illustrated in Fig. 3.21a–c. For the Sn–4Ag and Sn–37Pb solder joints, crack propagates inside the solder close to the solder/IMC interface, while fatigue cracks of the Sn–Bi/Cu joints propagate exactly along the interface. During the

Fig. 3.20 Evolution of tensile strength of the Sn–4Ag and Sn–58Bi solder alloy during the aging process. Reprinted from Ref. [11], Copyright 2010, with permission from Elsevier

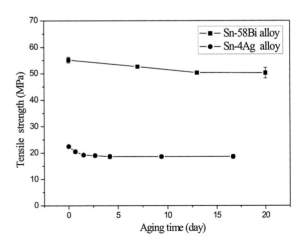

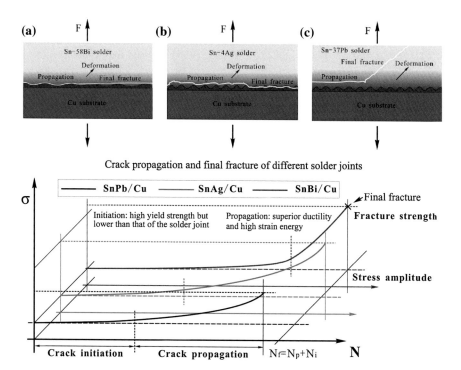

Crack propagation and final fracture of different solder joints

Fig. 3.21 The fatigue fracture mechanisms of Cu/solder joint interfaces and influencing factors of fatigue resistance: **a–c** fatigue crack propagation paths. Reprinted from Ref. [11], Copyright 2010, with permission from Elsevier

crack propagation process, the real applied stress of the solder joints increases with increasing cyclic number; when the real stress approaches to the fracture strength of solder joints, the final fracture occurs. The final fracture modes are similar to the solder joints fractured at high strain rate, affected by the mechanical properties of the solder and the interfacial IMC.

3.4 Tensile-Compress Fatigue Damage Mechanisms

Based on the discussion on fatigue fracture processes, the influencing factors of the fatigue resistance at the two stages are predicted. It is widely accepted that high ductility enhances the low-cycle fatigue resistance of a material [36], but the ductility is not the only influencing factor on fatigue resistance. The fatigue life of a typical fatigue fracture (N_f) consists of the crack initiation (N_i) and propagation cycles (N_p). For the solder joints, as the crack initiation process is induced by local plastic deformation of solder, the initiation cycles (N_i) are affected by the yield strength of solder. With higher yield strength, the solder is too rigid to deform and micro-cracks are difficult to initiate; thus the Sn–Bi/Cu solder joints should have higher N_i than the Sn–Ag/Cu and SnPb/Cu solder joints at low stress amplitude. For the Sn–Ag/Cu and Sn–Pb/Cu solder joints, crack propagates inside the solder; a plastic deformation region was formed at the tip of crack. Therefore, the deformation energy of the solder alloy, as the resistance to crack propagation, is the dominative factor to the crack propagation cycles. As the Sn–Ag solder has higher fatigue resistance than the Sn–Pb solder, the Sn–Ag/Cu solder joints have higher N_p than the Sn–Pb/Cu joints. Although the Sn–Bi solder has good ductility, it shows little plastic strain at the low stress amplitude. Observation on the fracture surface of the Sn–Bi/Cu joints has proved that the fatigue crack is propagated along the solder/IMC interface. Because the plastic deformation of the Sn–Bi solder near the interface is slight and the IMC cannot exhibit any plastic deformation, the resistance to crack propagation is low. Therefore, the Sn–Bi/Cu joint sample should have high propagation rate and low propagation cycles.

Figure 3.21 gives a simple description on the fracture process of the three solder joints and makes a comparison on their crack initiation, propagation rate, and fracture strength. The x-axis is the fatigue cycles and the y-axis is the real stress, and the three lines represent the real stress of the three different solder joints. The real stress increases from the stress amplitude to the fracture strength during the crack initiation and propagation process. As in the figure, the Sn–Pb/Cu joints have the lowest fracture strength and fatigue life at the same stress amplitude. The Sn–Bi/Cu solder joints have the highest fracture strength and N_i, but low N_p, because of their high crack propagate rate. The Sn–Ag/Cu joints have a balance distribution of N_i and N_p. It is well known that the fracture strength of solder joints is affected by the interfacial microstructure [22, 23, 37]. From Fig. 3.21, it is obvious that the propagation cycles also depend on the fracture strength, and thus the interfacial structure and the IMC layer can affect the fatigue life by dominating the fracture strength. In addition, for all of the three solder joints, the IMC thickness keeps increasing with increasing aging time, but the IMC layer never fractures during the crack initiation process; thus the IMC thickness may have little influence on crack initiation mechanism. Besides, though there is obvious coarsening of the solders during the aging process, the decrease in yield strength of them is very slight. Since the stress localization is induced by the deformation mismatch between the solder and IMC layer, the coarsening does not induce an essential transition on crack initiation mechanism though it may affect the crack propagation rate a little.

3.5 Brief Summary

In this chapter, the tensile-compress fatigue properties and damage behavior of the Sn–4Ag/Cu, Sn–58Bi/Cu, and Sn–37Pb/Cu solder joints are investigated; the fatigue life and interfacial fatigue damage mechanisms of the solder joints are compared; and the influencing factors on fatigue life were discussed. Based on the observation results and discussions, the following conclusions can be drawn:

1. The thickness of the IMC layer at the Sn–4Ag/Cu interface increases with increasing aging time, and the tensile strength of the solder joints keeps decreasing, accompanying the transform in fracture mechanism: from dimple fracture inside the solder to cleavage fracture inside the IMC layer. Since the decrease in strength of solder during the aging process is little, the serious decrease in tensile strength of the solder joint and the transform of fracture mechanism should be related to the increase of IMC thickness. The evolution in microstructure and tensile strength of the Sn–58Bi/Cu solder joints is similar to that of the Sn–4Ag/Cu solder joints, while after aging for a long time, the Bi segregation at the Cu_3Sn/Cu interface will result in an interfacial embrittlement, making the joint strength to decrease sharply.

2. The three groups of solder joints show similar stress amplitude–fatigue life relationship. At low stress amplitude, the tensile-compression fatigue life decreases exponentially with increasing stress amplitude. When the stress amplitude is close to the yield strength of the solder, the fatigue life shows an abnormal increase due to strain hardening of the solder. The fatigue life consists of the crack initiation and propagation cycles; fatigue lives of both the Sn–4Ag/Cu and the Sn–58Bi/Cu solders are higher than the Sn–37Pb/Cu solder joints.

3. Under cyclic tensile-compression loadings, obvious strain concentration occurs at the solder/IMC interface, and fatigue crack initiates along or around the interface. The fatigue crack in the Sn–4Ag/Cu and Sn–37Pb/Cu solder joint propagates inside the solder close to the joint interface, while the fatigue crack in the Sn–58Bi/Cu solder joint propagates along the solder/IMC interface. When the real stress increases to the interfacial fracture strength, a final fracture similar to the fracture at high strain rate occurs.

4. The ductility of solder can significantly affect the fatigue life of solder joint through influencing the crack initiation, propagation path, and propagation rate. The thickness of the interfacial IMC layer has little influence on crack initiation and propagation mechanisms, but it can affect the fatigue life by dominating the fracture strength of solder joints. The solder joints aged for different times have similar crack initiation process and initiation cycles, and only the difference in tensile strength leads to different crack propagation cycles and different life cycles.

References

1. Abtew M, Selvaduray G. Lead-free solders in microelectronics. Mater Sci Eng R. 2000;27:95–141.
2. Kim HK, Tu KN. Rate of consumption of Cu in soldering accompanied by ripening. Appl Phys Lett. 1995;67:2002–4.
3. Seah SKW, Wonga EH, Shim VPW. Fatigue crack propagation behavior of lead-free solder joints under high-strain-rate cyclic loading. Script Mater. 2008;59:1239–42.

4. Erinc M, Assman TM, Schreurs PJG, Geers MGD. Fatigue fracture of SnAgCu solder joints by microstructural modeling. Int J Fract. 2008;152:37–49.
5. Zhu QS, Zhang ZF, Shang JK, Wang ZG. Fatigue damage mechanisms of copper single crystal/Sn-Ag-Cu interfaces. Mater Sci Eng A. 2006;435–436:588–94.
6. Anderssona C, Lai Z, Liu J, Jiang H, Yu Y. Comparison of isothermal mechanical fatigue properties of lead-free solder joints and bulk solders. Mater Sci Eng A. 2005;394:20–7.
7. Zhao J, Mutoh Y, Miyashita Y, Mannan SL. Fatigue crack-growth behavior of Sn-Ag-Cu and Sn-Ag-Cu-Bi lead-free solders. J Electron Mater. 2002;31:879–86.
8. Kanchanomai C, Mutoh Y. Effect of temperature on isothermal low cycle fatigue properties of Sn-Ag eutectic solder. Mater Sci Eng A. 2004;381:113–20.
9. Kanchanomai C, Limtrakarn W, Mutoh Y. Fatigue crack growth behavior in Sn-Pb eutectic solder/copper joint under mode I loading. Mech Mater. 2005;37:1166–74.
10. Laurila T, Vuorinen V, Kivilahti JK. Interfacial reactions between lead-free solders and common base materials. Mater Sci Eng R. 2005;49:1–60.
11. Zhang QK, Zhu QS, Zou HF, Zhang ZF. Fatigue fracture mechanisms of Cu/lead-free solders interfaces. Mater Sci Eng A. 2010;527:1367–76.
12. Zhang QK, Zou HF, Zhang ZF. Tensile and fatigue behaviors of aged Cu/Sn-4Ag solder joints. J Electron Mater. 2009;38:852–9.
13. Zribi A, Clark A, Zavalij L, Borgesem D, Cotts EJ. The growth of intermetallic compounds at Sn-Ag-Cu solder/Cu and Sn-Ag-Cu solder/Ni interfaces and the associated evolution of the solder microstructure. J Electron Mater. 2001;30:1157–64.
14. Deng X, Piotrowski G, Williams JJ, Chawla N. Influence of initial morphology and thickness of Cu6Sn5 and Cu3Sn intermetallics on growth and evolution during thermal aging of Sn-Ag solder/Cu joints. J Electron Mater. 2003;32:1403–13.
15. Yang W, Messier RW, Felton LE. Microstructure evolution of eutectic Sn-Ag solder joints. J Electron Mater. 1994;23:765–72.
16. Ma X, Wang FJ, Qian YY, Yoshida F. Development of Cu-Sn intermetallic compound at Pb-free solder/Cu joint interface. Mater Lett. 2003;57:3361–5.
17. Bonda NR, Noyan IC. Effect of the specimen size in predicting the mechanical properties of PbSn solder alloys. IEEE Trans Compon Packag Manuf Technol A. 1990;19:208–12.
18. Yu SP, Hon MH, Wang MC. The adhesive strength of a lead free solder hot-dipped on Cu substrate. J Electron Mater. 2000;29:237–43.
19. Dao M, Chollacoop N, Van Vliet KJ, Venkatesh TA, Suresh S. Computational modeling of the forward and reverse problems in instrumented sharp indentation. Acta Mater. 2001;49:3899–918.
20. Deng X, Chawla N, Chawla KK, Koopman M. Deformation behavior of (Cu, Ag)-Sn intermetallics by nanoindentation. Acta Mater. 2004;52:4291–303.
21. Mei Z, Sunwoo AJ, Morris JW Jr. Analysis of low-temperature intermetallic growth in copper-tin diffusion couples. Metall Trans A. 1992;23:857–64.
22. Lee HT, Chen MH, Jao HM, Liao TL. Influence of interfacial intermetallic compound on fracture behavior of solder joints. Mater Sci Eng A. 2003;358:134–41.
23. Vaynman S, Fine ME, Jeannotte DA. Isothermal fatigue of low tin lead based solder. Metall Trans A. 1988;19:1051–9.
24. Glazer J. Metallurgy of low temperature Pb-free solders for electronic assembly. Int Mater Rev. 1995;40:65–93.
25. Ding Y, Wang CQ, Tian YH, Li MY. Influence of aging treatment on deformation behavior of 96.5Sn3.5Ag lead-free solder alloy during in situ tensile tests. J Alloys Compd. 2007;428:274–85.
26. Suna P, Andersson C, Wei XC, Cheng ZN, Shangguan DK, Liu JH. Study of interfacial reactions in Sn-3.5Ag-3.0Bi and Sn-8.0Zn-3.0Bi sandwich structure solder joint with Ni(P)/Cu metallization on Cu substrate. J Alloys Compd. 2007;437:169–79.
27. Zhao J, Miyashita Y, Mutoh Y. Fatigue crack growth behavior of 96.5Sn-3.5Ag lead-free solder. Int J Fatigue. 2001;23:723–31.

28. Liu PL, Shang JK. Interfacial segregation of bismuth in copper/tin-bismuth solder interconnect. Script Mater. 2001;44:1019–23.
29. Liu PL, Shang JK. Interfacial embrittlement by bismuth segregation in copper/tin–bismuth Pb-free solder interconnect. J Mater Res. 2001;16:1651–9.
30. Zhu QS, Zhang ZF, Wang ZG, Shang JK. Inhibition of interfacial embrittlement at SnBi/Cu single crystal by electrodeposited Ag film. J Mater Res. 2008;23:78–82.
31. Zou HF, Zhang QK, Zhang ZF. Eliminating interfacial segregation and embrittlement of bismuth in SnBi/Cu joint by alloying Cu substrate. Script Mater. 2009;61:308–11.
32. Keast VJ, Fontaine AL, Plessis JD. Variability in the segregation of bismuth between grain boundaries in copper. Acta Mater. 2007;55:5149–55.
33. Zhang QK, Zou HF, Zhang ZF. Improving tensile and fatigue properties of Sn-58Bi/Cu solder joints through alloying substrate. J Mater Res. 2010;25:303–14.
34. Pang XY, Shang PJ, Wang SQ. Weakening of the Cu/Cu3Sn(100) Interface by Bi Impurities. J Electron Mater. 2010;39:1277–82.
35. Mei Z, Morris JW Jr. Characterization of eutectic Sn-Bi solder joints. J Electron Mater. 1992;21:599–607.
36. Glazer J. Microstructure and mechanical properties of Pb-free solder alloys for low-cost electronic assembly. A review. J Electron Mater. 1994;23:693–700.
37. Kerr M, Chawla N. Creep deformation behavior of Sn-3.5Ag solder/Cu couple at small length scales. Acta Mater. 2004;52:4527–35.

Chapter 4
Shear Creep-Fatigue Behavior
of Cu/Pb-Free Solder Joints

4.1 Introduction

In the electronic component, the primary strain subjected by the solder joints is resulted from the difference in the CTEs between the chip, chip carrier, and the circuit board. Figure 4.1 illustrates the temperature curves and the corresponding thermal deformation in the solder joints [1]. When the component is heated from the equilibrium temperature, the solder joints will suffer positive shear strain, and the strain is reversed when it is cooled. Since the electronic equipments placed in service are periodically turned on and off, cyclic loading occurs in the solder joints and fatigue damage is induced. Consequently, the severity of the cyclic loading is a major factor in deciding the life cycle of the equipments [2].

During the thermomechanical cycling, the plastic deformation is apparently one of the major deformation mechanisms of the solders. Besides, creep deformation should also play an important role because the homologous temperature (T/T_m) of the solder used in electronic assembly usually exceeds 0.6. Under that condition, the microstructure of solder is unstable, creep and stress relaxation occur rapidly during the holding period at the high temperature within each cycle [2, 3]. As a result, the damage mechanism of the solder joints is a complex interaction between creep and low-cycle fatigue.

Since the creep-fatigue damage arise from different CTEs is one of the most probable failure modes of the solder joints [2, 4], and the corresponding fracture mechanism is crucially important for evaluating the mechanical reliability of the solder joints, thus far many related studies have focused on the creep behaviors of the solder and solder joints [5–10]. The creep stress exponents and activation energies of a series of Sn-based solders have been obtained, and two major creep deformation mechanisms of the solders are also proposed [4–8, 11]. However, investigations on creep-fatigue behavior of the solder joints are still lacking although it is more essential for the reliability evaluation. Besides, most of the current studies

© Springer-Verlag Berlin Heidelberg 2016
Q. Zhang, *Investigations on Microstructure and Mechanical Properties of the Cu/Pb-free Solder Joint Interfaces*, Springer Theses,
DOI 10.1007/978-3-662-48823-2_4

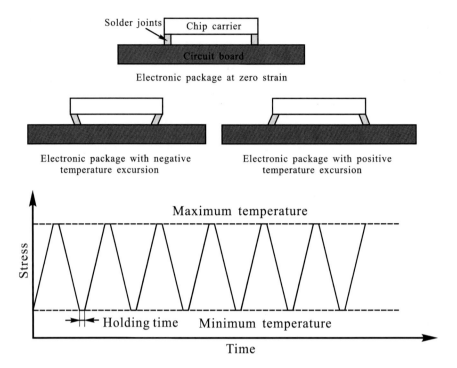

Fig. 4.1 Illustrations on thermal deformation and temperature curves of solder joints in electronic component. Reprinted from Ref. [1], Copyright 2011, with permission from Elsevier

only emphasis on the creep curves for measuring the creep parameters, while the visualized observations on the failure processes are less sufficient.

The discussion above indicates that a comprehensive, intuitionistic under-standing on creep-fatigue mechanisms of the lead-free solder joints is required. Therefore, in the present study the authors have concentrated their efforts on revealing the creep-fatigue behaviors of the representative Sn–Ag/Cu and Sn–Bi/Cu solder joints. The SEM equipped with in situ tensile stage and EBSD system was used to conduct the creep-fatigue tests, which allows in situ observation on the creep-fatigue process and characterizing the evolution of microstructure. The in situ observation can get visualized images on damage processes and has been widely used to investigate the deformation behaviors of solder joints [12, 13]. The joint interfaces in the employed specimens are relatively small in length scales (1 × 1 mm), in order to make them more similar to the solder joints in the electronic component. By analyzing the creep-fatigue curves, the deformation morphologies and evolution in microstructure, understandings on creep-fatigue behaviors of the solder joints are provided.

4.2 Experimental Procedure

The Sn–4Ag/Cu and Sn–58Bi/Cu solder joints were employed as example in this study, because the Sn–Ag series alloy is the most promising lead-free solder candidate [14], and Sn–58Bi is close in melting point and microstructure to Sn–37Pb solder. The cold-drawn oxygen-free high conductivity (OFHC) Cu clava with a purity of about 99.9 % was chosen as the substrate material, its yield strength is about 300 MPa. The Sn–4Ag solder alloy was prepared by melting high purity (>99.99 %) tin and silver at 800 °C for 30 min in vacuum. The Cu clava was first spark cut into small blocks with a step at one end, and then the surfaces at the end were ground and electrolytically polished for soldering. After air drying, a soldering flux was dispersed on the polished area to enhance wetting and minimize oxide formation. The steps of two Cu blocks were butt to butt, a solder alloy sheet was sandwiched between them and graphite plates were clamped on their sides to avoid the outflow of the molten solder. The prepared samples were put in an oven with a constant temperature (260 °C for Sn–4Ag/Cu solder joint and 200 °C for Sn–Bi/Cu solder joint), kept for 8 min after the melting of the solder and then cooled down in air. After that the joint samples were sliced into specimens by spark cutting; the side surfaces of the specimens were first ground with 2000# SiC abrasive paper and then carefully polished for the interfacial microstructure observations. The preparation process and dimension of the test specimens are illustrated in Fig. 4.2a.

The creep-fatigue tests were conducted by Gatan MTEST2000ES Tensile Stage equipped on the ZEISS Supra 35 field emission SEM (see Fig. 4.2b). Shear tests of the solder joints were performed first to find proper conditions for the creep-fatigue tests, the crossbeam speed was set as 0.05 mm min^{-1}. The shear strain rate is calculated by dividing the crossbeam speed with the solder thickness. To reveal the creep behaviors of the solder during the holding period, the shear tests were paused at some certain stresses to show the stress relaxation. All the creep-fatigue testes were stress-controlled, the shear stress ranges were chosen to be 2–23, 10–25, or 2–25 MPa for the Sn–4Ag/Cu solder joints, and 2–20, 2–22 MPa for the Sn–58Bi/Cu solder joints, and the holding times were set as 30 or 120 s. The test conditions were expressed in the form of "stress range/holding time." The tests were set as tension–tension mode because the working temperature of the electronic component is usually higher than the equilibrium temperature. During the creep-fatigue process, the force and displacement were recorded automatically by the tensile stage. The full views of the solder joints deformed for different cycles were first observed to show the macroscopic damage process, and the interfacial deformation morphologies were also tracked with emphasis as there is usually strain concentration at the joint interface. In order to reveal the deformation mechanisms of the solder, the creep-fatigue tests of selected solder joints were paused after certain cycles and the microstructures of the solder were characterized by EBSD. After the tests, the fracture morphologies of the specimens were observed by SEM.

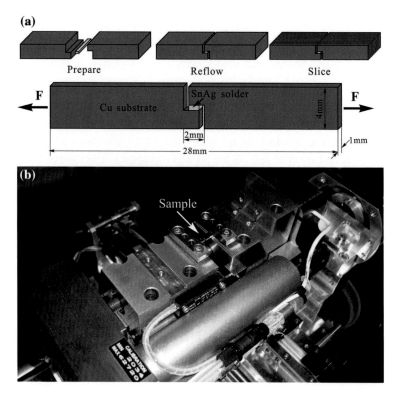

Fig. 4.2 **a** Preparation process and dimension of test specimens; **b** image of the Gatan MTEST2000ES in situ tensile stage. Reprinted from Ref. [1], Copyright 2011, with permission from Elsevier

4.3 Shear and Creep-Fatigue Behavior of Sn–4Ag/Cu Solder Joint

4.3.1 Shear and Stress Relaxation Behavior

Figure 4.3 presents the shear and stress relaxation behaviors of the Sn–4Ag/Cu solder joints. The shear stress–strain and stress–time curves are shown in Fig. 4.3a, b, respectively. The shear strain is calculated by first subtracting the elastic displacement of Cu from the total displacement, and then divided the result with the solder thickness. It can be found that the shear yield strength of the solder joint is about 24 MPa and the shear fracture strength is about 32 MPa, while obvious stress relaxation occurs when the test was paused at the stresses higher than 20 MPa. During the holding period, the solder is similar to be deformed at a very low strain rate, and the strain is contributed by plastic deformation and creep [3]. The creep deformation can generate a decrease in stress, i.e., the stress relaxation. Besides, as the yield strength of solder is lower at lower strain rate [15], the stress decreases

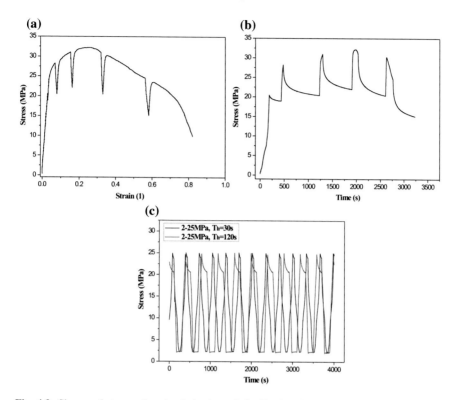

Fig. 4.3 Shear and stress relaxation behaviors of the Sn–4Ag/Cu solder joints: **a** shear stress–strain curve, **b** shear stress–time curve, **c** stress–time curve during the creep-fatigue test at the conditions of 2–25 and 2–25 MPa/120 s. Reprinted from Ref. [1], Copyright 2011, with permission from Elsevier

when the test is paused. The stress relaxation behavior is affected by the holding stress and time. As shown in Fig. 4.3a, b, the stress relaxation rate is higher at higher holding stress, and both the stress and the relaxation rate decrease with increasing holding time. Similar stress relaxation behaviors have also been observed in the Sn–3.5Ag and Sn–9Zn solder alloys [16], which suggests some validity for characterizing the dependence of strength on time as an indicator of the creep behaviors. According to the shear strength and stress relaxation behaviors, the peak stresses (the upper limit) of the stress ranges were chosen to be 23 or 25 MPa and the holding time was set as 30 or 120 s to get proper lifetimes. Under that conditions, the Cu substrate only shows a little elastic deformation, thus the strain of solder joints is contributed by deformation of solder. Figure 4.3c shows the stress–time curve of the creep-fatigue test at 2–25 MPa/120 s, in which obvious stress relaxation occurs during the holding period at 25 MPa, and the relaxation rate has become very low after 120 s. As the peak stresses approximate the yield strength of the solder, not only classical creep but also plastic deformation occurs during the holding period.

4.3.2 Strain–Time Relationship During Creep-Fatigue Process

For all the solder joints, the creep-fatigue processes are similar in general. Figure 4.4 shows the strain–time relationship of a specimen tested at the condition of 2–25 MPa/120 s as an example. Similar to the typical creep curves of the metallic materials, the low-cycle creep-fatigue process can also be divided into three stages according to the increasing rate of strain. During the initial few cycles, the strain increases rapidly but its increment per cycle decreases gradually, which should be resulted from the increasing strain hardening rate of the solder [17, 18]. After a few cycles, the strain starts to increase linearly with increasing cycles, indicating that the creep-fatigue process has entered a steady stage. The second stage comprises most of the life cycles. When the strain increases to a certain value, the creep-fatigue failure process accelerates gradually, and the final fracture occurs after the strain increases sharply in the lattermost few cycles. According to their features, the three stages are defined as strain hardening stage, steady deforming stage, and accelerating fracture stage, respectively.

The strain–time relationships of the solder joints tested at different conditions are gathered in Fig. 4.5 to make a comparison. As in the figure, the solder joint tested at higher peak stress fractures in a higher rate and has a lower lifetime. At the same stress range, longer holding time correlates with lower lifetime, because the deformation keeps occurring during the holding period at the peak stress. For the solder joint tested at 10–25 MPa/30 s, the increasing rate of strain is also higher since its duration time at the high stress is longer. All the solder joints with the peak stress of 25 MPa have similar fracture strains, even the average stresses are different, while that tested at 2–23 MPa has a higher fracture strain. Although most of the curves have the aforesaid three stages, it is notable that the increasing rate of strain of the solder joint tested at 2–23 MPa increases gradually at the second half of the creep-fatigue process, in other words, its accelerating fracture stage starts earlier

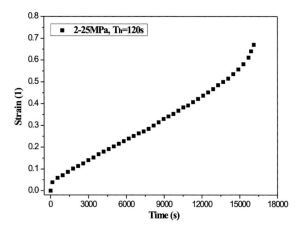

Fig. 4.4 Strain–time curve of the specimen tested at 2–25 MPa/120 s. Reprinted from Ref. [1], Copyright 2011, with permission from Elsevier

Fig. 4.5 Strain–time curves of the specimens tested under different conditions. Reprinted from Ref. [1], Copyright 2011, with permission from Elsevier

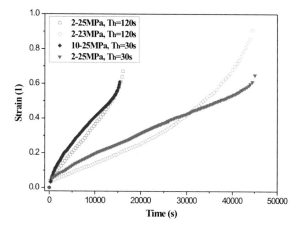

Fig. 4.6 Precise strain–time curves of different specimens at the steady deforming stage. Reprinted from Ref. [1], Copyright 2011, with permission from Elsevier

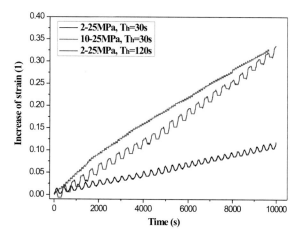

than the other solder joints. Besides, its lifetime is also lower than expected. Therefore, its fracture processes will be exhibited later to reveal the reasons.

Precise strain–time curves in the steady deforming stage are shown in Fig. 4.6. For each specimen, the strain–time curves of each cycle are very similar: the strain increases when the solder joint is in tension and declines when the stress decreases, and the resultant increment in strain per cycle is constant. In contrast, the deformation behaviors of the solder joints tested at different conditions are quite different. For the solder joint tested at 10–25 MPa, the strain amplitude is much lower, and the cycle time is shorter. As a result, it has higher life cycles, while its lifetime is the lowest since its duration time at higher stress is longer. On account of the same reason, when the stress range is the same, the increasing rate of strain for the solder joints with longer holding time is much higher. Because the deformation behaviors of the solder joints are similar within each cycle, the strain increases linearly with increasing cycles in the second stage.

4.3.3 Deformation and Fracture Behavior of Solder Joint

The macroscopic deformation features of the solder joint tested at 2–25 MPa/30 s for different cycles are shown in Fig. 4.7, the strain and time (second) are tagged in each figure. The nine figures can be divided into three groups to correlate with the three stages. In the first stage, the deformation of solder is not obvious, as in Fig. 4.7a, b, only some slight plastic deformation occurs around the solder/Cu interface. Figure 4.7c–g shows the deformation morphologies at the second stage, in which the deformation of solder becomes more and more serious with increasing cycles. The deformation around the joint interface is particularly serious, and some microcracks initiate there. Due to the strain concentration around the large plate-like Ag_3Sn in the solder alloy [19], there are also some microcracks inside the solder, but they show no further development. In the third stage, the interfacial microcracks connect and evolve into long cracks, making the solder joints unstable, as in Fig. 4.7h, i, the final fracture occurs a few cycles after that.

The magnified morphologies at the joint interface deformed for different cycles were observed to reveal the crack initiation behaviors and are exhibited in Fig. 4.8. After 30 cycles, serious plastic deformation has occurred inside the solder close to the interface, as in Fig. 4.8a, while the deformation of solder a little far from the interface is very little, implying that there is strain concentration close to the interface. Figure 4.8b presents the interfacial morphologies after 70 cycles. The plastic deformation of solder close to the interface becomes more serious, and some wavy lines appear at the surface. Since the deformation of the solder with different

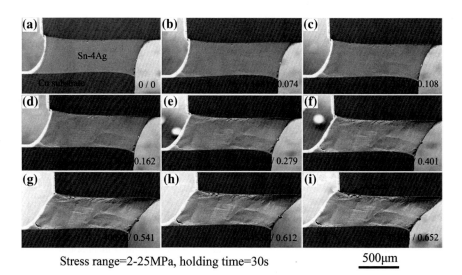

Stress range=2-25MPa, holding time=30s   500μm

Fig. 4.7 Macroscopic morphologies of specimen tested at 2–25 MPa/30 s for **a** 0, **b** 5, **c** 10, **d** 20, **e** 50, **f** 80, **g** 120, **h** 131, **i** 132 cycles. Reprinted from Ref. [1], Copyright 2011, with permission from Elsevier

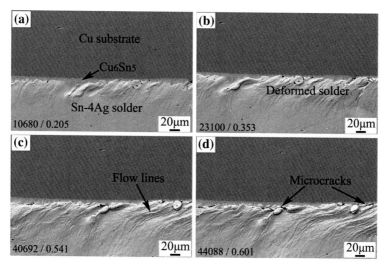

Fig. 4.8 Interfacial morphologies of specimen tested at 2–25 MPa/30 s for **a** 30, **b** 70, **c** 120, and **d** 130 cycles. Reprinted from Ref. [1], Copyright 2011, with permission from Elsevier

distances from the interface is not uniform, there is relative displacement between them, leading to the formation of the "flow lines." After 120 cycles, even the deformation of the solder a little bit far from the interface becomes serious, and the flow lines are more obvious (see Fig. 4.8c). Prior to the final fracture, there are some microcracks appearing at the Cu_6Sn_5/solder interface, as exhibited in Fig. 4.8d. Though the location in Fig. 4.8 is not the position where the major crack initiates, it can still reflect the interfacial damage and crack initiation behaviors during the creep-fatigue process.

Figure 4.9 shows the macroscopic morphologies of the solder joint tested at 2-23 MPa/120 s. As its strain–time curve is a bit abnormal (see Fig. 4.5), it is necessary to explore its failure process. At the early stage, its evolution is similar to that in Fig. 4.7, as in Fig. 4.9a–d. After about half of its lifetime, a microcrack appears at the corner of this solder joint (see Fig. 4.9e). In Fig. 4.9f, the crack has become remarkable, and it appears to be a sliding fracture between the solder and Cu substrate. During the subsequent process, the sliding distance increases rapidly, as presented in Fig. 4.9f–h. As the sliding fracture can significantly decrease the area of the solder joint to support load and increase the real shear stress, the creep-fatigue failure process accelerates gradually with propagation of the sliding crack. When the real stress reaches the shear fracture strength of the specimen, final fracture will occur (see Fig. 4.9i). Although the deformation behaviors of the solder joints in Figs. 4.7 and 4.9 are similar in general, the primary cracks initiated at different locations, inducing different fracture behaviors in the latter process. Since the corner is usually a preferential location of stress concentration [20], if some microcracks initiate there, they can easily evolve into a sliding fracture and obviously accelerates the failure process. Figure 4.10 shows the enlarged views of the

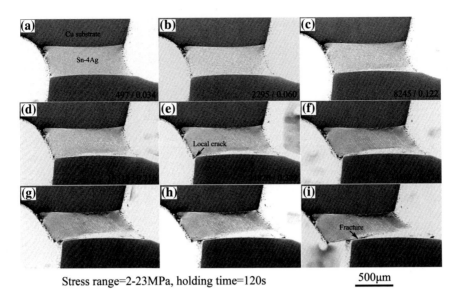

Stress range=2–23MPa, holding time=120s   500μm

Fig. 4.9 Macroscopic morphologies of specimen tested at 2–23 MPa/120 s for **a** 1, **b** 5, **c** 18, **d** 36, **e** 54, **f** 75, **g** 85, **h** 90, **i** 95 cycles. Reprinted from Ref. [1], Copyright 2011, with permission from Elsevier

sliding crack at the corner. As in Fig. 4.10a, the distribution of deformation is similar to that in Fig. 4.8, i.e., the plastic deformation of the solder near the joint interface is serious. After the sliding crack occurs, the strain concentration at the tip of the crack becomes more obvious (see Fig. 4.10b–d). Therefore, the sliding crack develops rapidly with increasing strain and accelerates the creep-fatigue failure process.

To reveal the final fracture behavior, the fracture surfaces of the solder joints were observed and are shown in Fig. 4.11. Figure 4.11a exhibits the side surface of the solder joint prior to its final fracture. According to its feature, it can be predicted that the final fracture occurs around the joint interface. The fracture surfaces of the two sides are a little bit different. As in Fig. 4.11b, some dimples of different sizes are evident on the surface of side 1 (the solder side). In contrast, the fracture surface of the opposite side has less dimples and consists of two regions with different features (see Fig. 4.11c), indicating that there is a transition in fracture mode during the fracture process. Figure 4.11d, e shows the appearance of region 2 at the two sides. As the elongated dimples are very similar to those at the shear fracture surfaces [21, 22], the fracture in region 2 is considered as a classical ductile fracture: the microcracks along the Cu_6Sn_5/solder interface were elongated, forming the dimples on the fracture surface. In contrast, not only the solder but also abundant cracked Cu_6Sn_5 grains were observed in region 1, as in Fig. 4.11f. Because the Sn–Ag/Cu solder joint is more apt to fracture inside the interfacial IMC layer at higher stress and strain rate [22–24], the fracture in region 1 should be the ultimate

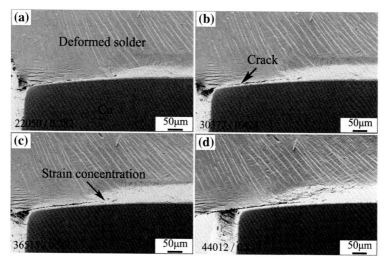

Fig. 4.10 Cracking at the corner of specimen tested at 2-23 MPa/120 s for **a** 48, **b** 66, **c** 79, and **d** 95 cycles. Reprinted from Ref. [1], Copyright 2011, with permission from Elsevier

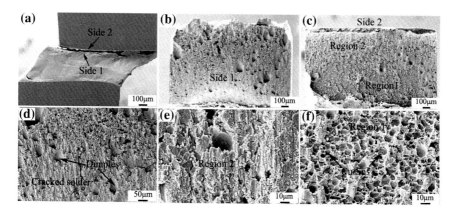

Fig. 4.11 Fracture morphologies of Sn–4Ag/Cu solder joints: **a** side surface prior to the final fracture, **b** and **c** macroscopic fracture surfaces, **d, e,** and **f** microscopic fracture surfaces. Reprinted from Ref. [1], Copyright 2011, with permission from Elsevier

fracture. The micrographs of fracture surfaces in Fig. 4.11 suggest that the final failure starts inside the solder in a ductile fracture mode, and the ultimate fracture occurs inside the interfacial Cu_6Sn_5 layer.

4.3.4   Evolution in Microstructure of Solder

The microstructures of the Sn–4Ag solder in the specimen tested at 2–25 MPa/120 s
for different cycles were characterized by EBSD to reveal the deformation mechanisms of solder. The periodic time of each cycle is 400 s and time for each EBSD
characterization is about 2 h. Figure 4.12 presents strain–cycle relationship and the
corresponding grain maps of the solder. As the creep-fatigue test is not continuous,
the strain–cycle relationship rather than the strain–time relationship is exhibited in
the figure, and the cycles of the grain maps are indicated by the arrows. The
different colors correspond to the solder grains with different orientations (see the
orientation triangle in the figure), the low-angle grain boundaries ($2° < \Delta\theta < 15°$)
are indicated with white lines and the high-angle boundaries ($\Delta\theta > 15°$) are indicated with black lines. Before the test, the solder consists of coarse grains with four
dominant orientations and are separated by large-angle grain boundaries, as shown
in Fig. 4.12a. The grains are named by their colors to distinguish and describe them.
It is interesting to find that the "yellow" grains are not continuous but have the same
orientation. In fact, as the specimen is sliced from a large solder joint, the yellow
grains are actually fractions of the same solder grain, thus their orientations are
exactly the same. After a few cycles, the "yellow" grains become much thinner (see
Fig. 4.12b), implying that grain-boundary migration occurs between the "yellow"
and the "purple" grains. There are two major influencing factors on grain-boundary
migration, i.e., the strain energy and the boundary curvature. Since there is obvious

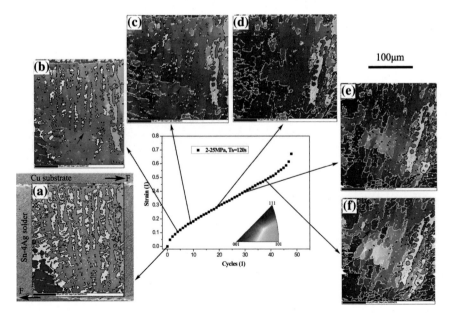

Fig. 4.12 Grain maps of Sn–4Ag solder deformed at 2–25 MPa/120 s for: **a** 0, **b** 3, **c** 8, **d** 18, **e** 29,
f 37 cycles. Reprinted from Ref. [1], Copyright 2011, with permission from Elsevier

plastic deformation inside the solder after a few cycles, the strain energy can be a driving force for grain-boundary migration. On the other hand, the "yellow" grains should have less than six convex boundaries as they are fractions of the large grain, thus they are likely to be absorbed by the large grains through grain-boundary migration. Besides, sight color difference appears at different locations of the "purple" grain, and a few discontinuous low-angle grain boundaries were also observed. In Fig. 4.12c, the color difference becomes more obvious and longer grain boundaries emerge between the fractions with color difference (see the red arrows), which results in subdivision of the original solder grains. Similar subdivisions have also been observed in the Sn–3.5Ag solder and Al alloy undergoes serious plastic deformation [17, 25], and the dislocation movement and rearrangement are considered to be the mechanisms. Grain rotation and grain-boundary sliding are predicated to occur in the newly formed grains during the further deformation process, because their orientations change obviously. The evolutions in shape and orientation of the grain indicated by the blue arrows provide a support for that predication. Due to the rotation, the misorientations between the solder grains become larger with increasing strain, and new grain boundaries are formed continuously to further subdivide them into finer grains (see Fig. 4.12d–f). Since the grain-boundary migration did not occur between the "green" and "purple" solder grains, the images of the former can clearly show the deformation of solder. The mechanisms of the solder grain subdivision and its influences on the creep behavior will be discussed later in more detail.

4.3.5 Creep-Fatigue Mechanisms of Sn–4Ag/Cu Solder Joint

Based on the results provided above, the creep-fatigue process can be illustrated in Fig. 4.13 with a strain–cycle curve. The whole process is divided into three stages according to the increasing rate of strain. During the initial few cycles, the strain increases in a high rate as the solder is very soft. With the deformation processing, the strain hardening of solder develops continuously, making it more and more rigid. After certain cycles, the strain hardening reaches the saturated state and the increasing rate of strain becomes constant. As the stress ranges are very high in this test, the strain hardening stages are transient. According to the slope of the strain–time curves, there are about five cycles, and the corresponding saturated shear strain is about 4–5 %. Slight deformation inside the solder can be observed after the first stage, but no cracking occurs.

After the strain hardening reaches a saturated state, the solder joints behave in a constant deformation, and the creep-fatigue process enters a steady stage. Since the deformation behavior of solder and the sectional area of the solder joints change little, the deformation curve within each cycle is similar and the strain increases

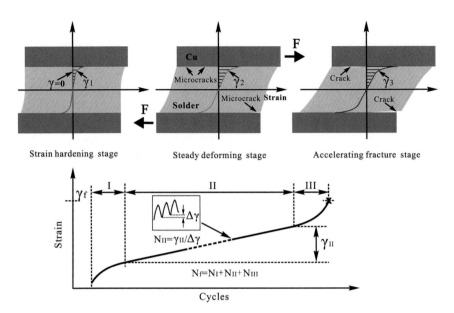

Fig. 4.13 Illustrations on creep-fatigue failure process of the Sn–4Ag/Cu solder joints. Reprinted from Ref. [1], Copyright 2011, with permission from Elsevier

linearly with increasing cycles. The cycles in stage II can be calculated by the following equation:

$$N_{II} = \gamma_{II}/\Delta\gamma \tag{4.1}$$

where N_{II} is the cycles in stage II, γ_{II} is the total shear strain in stage II and $\Delta\gamma$ is the increment in shear strain per cycle. On the basis of the equation, one can calculate the lifetime and cycles from γ_{II} and $\Delta\gamma$. The increment in strain per cycle $\Delta\gamma$ is dominated by the test conditions and can be estimated after the creep-fatigue process enters the second stage. The total strain in stage II γ_{II} has a proportional relationship with the creep-fatigue fracture strain. As in Fig. 4.5, its proportion in the fracture strain is 80–90 %. Therefore, the cycles in stage II can be estimated provided the creep-fatigue fracture strain is known. As the second stage comprises most of the lifetime, the whole fatigue life can also be approximately estimated.

During the second stage, there are some microcracks initiated at the solder/Cu_6Sn_5 interface. When the microcracks connect to form long cracks, local shear fractures occur at the interface. These fractures can significantly degrade the capability of the solder joint to support loads, resulting in obvious increase in the damage rate. As a symptom, it signifies that the creep-fatigue process enters the accelerating fracture stage. In this stage, the strain is not only contributed by deformation of solder, but also by the relative sliding between the solder and the substrate. The sliding can badly degrade the adhesive property of the joint interface, making the real stress increase sharply with increasing strain. When the real stress increases to the shear fracture strength, final fracture occurs in the loading phase of the cycle. The final fracture condition is:

$$\tau_r = \tau_{max}/(1-\Delta s) = \tau_f \quad\quad\quad (4.2)$$

where τ_{max} is the peak stress, τ_r is the real shear stress, τ_f is the shear fracture strength of the solder joint, and Δs is a parameter describing the decreasing ratio in shear fracture strength of the specimen, which relates to the shear fracture strain. As τ_{max} is set to be 23 or 25 MPa, and τ_f is 32 MPa, Δs is calculated to be 0.22 or 0.28. If the whole creep-fatigue displacement is divided with the length of the joint interface, it can be found that for all the specimens, the results are notably approximated with Δs. Therefore, the creep-fatigue fracture strain can be experimentally estimated from the shear strength and the stress range. In consequence, the creep-fatigue life can be approximately estimated as discussed before.

The deformation and fracture behaviors of the solder joints are illustrated in Fig. 4.13 with a two-dimensional coordinate system, the center of the solder joint is chosen as the reference point, the vertical horizontal axis indicates the position, and the horizontal axis indicates the displacement from the original position. The displacements of the solder mass points at different strains are illustrated by four lines. At the initial state, the shear strain is zero and the displacement distribution line is overlapped with the vertical horizontal axis (see the green line). The three bending curves show the displacement distributions at three different strains ($\gamma_1 < \gamma_2 < \gamma_3$). For the solder a little bit far from the joint interface, the displacement is proportional to its distance from the reference point, while the displacement of solder close to the interface is notably higher because of the interfacial strain concentration. With increasing strain, the displacements at each position increases, while the distribution characteristics changes little. Microcracks initiate along the interface due to deformation mismatch between the solder and the Cu_6Sn_5 layer [24], but they should not significantly decrease the capacity of the solder joints to support load before evolving into long cracks, otherwise the increasing rate of strain will increase obviously. Whereas, if the microcracks initiate at the corner of the joints, they are easier to develop into sliding cracks and accelerate the fracture process. In the real service condition of the electronic components, the stress endured by the solder joints is much lower than that in the test, but the general fracture process should be similar.

During the creep-fatigue process, the total inelastic strain of the solder joints is contributed by plastic deformation and creep deformation of the solder. Accordingly, it is necessary to explore the corresponding deformation mechanisms. The grain maps in Fig. 4.12 have demonstrated that the grain subdivision and grain rotation occur inside the solder after the strain hardening becomes saturated. The subdivision process usually consists of two steps, i.e., formation of subgrains and subsequently the increase in the misorientations. At the early stage of the creep-fatigue process, the solder deforms in the classical dislocation slip and multiplication mechanisms, and the dislocation density keeps increasing with that process. When the density reaches a certain extent, the dislocation tangle and intersect will make them difficult to slip, and the strain hardening becomes saturated. However, a high density of dislocations provides a necessary condition for the grain subdivision, because the low-angle grain boundary is composed of a group of edge dislocations. On the basis of the high dislocation density of dislocations and the high homologous temperature ($\sim 0.6T_m$), the high temperature recovery, i.e., polygonization occurs inside the solder. Through thermally activated dislocation movement and rearrangement [25–27], subgrain boundaries are formed. The zones with slight color difference in Fig. 4.12b are actually separated by subgrain boundaries with the misorientations angle less than 2°. As a primary mechanism for accommodating plastic deformation [25], the subgrain rotation occurs in the solder with increasing strain, resulting in continuous increment in misorientation angles and the transition from subgrain boundaries to low-angle grain boundaries. The subdivision and subgrain rotation keeps occurring with the strain continued, more grain boundaries are formed, and the subdivision taking place on a finer scale. In general, the grain subdivision and grain rotation are the major plastic deformation mechanisms of the solder.

It has been well accepted that dislocation climb and grain-boundary sliding are two major creep deformation mechanisms for both the Pb-bearing and the Pb-free solder alloys [6–8], and their contributions depend on the strain rate, stress level, and temperatures [28]. Dislocation climb is the dominant mechanism at higher strain rate, higher stress level, and lower temperature, while grain-boundary sliding is more expected at higher homologous temperature and lower strain rate [2]. Therefore, dislocation climb may show a greater dominance in the creep deformation of Sn–Ag and Sn–Ag–Cu solders because of their lower homologous temperatures compared with the SnPb solders. In fact, Shine et al. [29] have proposed that cyclic fatigue life correlates primarily with dislocation climb; experiment results of Mavoori et al. [30] also indicate that dislocation climb dominates the creep deformations of Sn–Ag alloy at the higher strain rates. Based on that, the dislocation climb is predicated to be the major creep mechanisms of solder in this study, especially at the early stage, because the original solder is composed by coarsen grains and the grain boundaries are rather lacking. After the grain subdivision, however, grain-boundary sliding may be promoted since the grain-boundary density becomes much higher. As the strain rate is very low during the holding period, grain-boundary sliding is predicated to occur between the thin grains and contributes a higher proportion in the creep deformation. The changes in shapes of some thin grains provide a support for that.

Based on the understandings on the creep-fatigue mechanisms, some intrinsic and extrinsic factors can affect the creep-fatigue behaviors. The influences of stress range, mean stress, and holding time are evident and have been shown before. The other factors include shape, dimension, and microstructure of the solder joints. Since the mechanical property of solder is dominated by its microstructure [31], and the strain of the solder joints is contributed by deformations of the solder, the microstructure of the solder has significant influence on the creep-fatigue behaviors. The solders composed by fine grains have higher yield strength and therefore higher resistance to plastic deformation, because the strength of solder is primarily affected by its grain size. Although the thinner grain size may promote the grain-boundary sliding, the latter is not the major creep mechanism. The interfacial microstructure of solder joints can significantly affect the crack initiation and final fracture behaviors. As the crack initiation is induced by plastic deformation and deformation mismatch at the joint interface, a higher strength of the solder means higher resistance to the crack initiation [24]. Besides, the final fracture condition of the specimen depends on the shear fracture strength, and the latter is significantly affected by the interfacial IMC thickness [32, 33]. In consequence, the creep-fatigue life is also influenced by the IMC thickness. With increasing interfacial IMC thickness, the shear strength of the solder joint decreases and the final fracture occurs earlier, resulting in a lower creep-fatigue life. The shape and dimension of the solder joints are major external factors in deciding the creep-fatigue behavior. For example, when the distance between the solder joints is longer or the size of joint is smaller, the strain and stress applied on the joint will be higher, and its creep-fatigue resistance is lower. The shape of the solder joints has significant influence on the strain localization behavior [20], in turn it can affect the crack initiation process because the cracks usually initiate at the strain concentration region. If there is some serious stress concentration region in the solder joint, the cracking will be easier to occur. In addition, some defects can affect the creep-fatigue life because they are usually the crack sources, especially when they are located at the strain concentration region.

4.4 Shear and Creep-Fatigue Behavior of Sn–58Bi/Cu Solder Joint

4.4.1 Shear Behavior of Sn–58Bi/Cu Solder Joint

Figure 4.14 exhibits the shear stress–strain and stress–time relationships of the Sn–58Bi/Cu solder joints. The shear strains of the solder joints were calculated by first subtracting the displacement of Cu from the total displacement, and then divided the result with the length of the solder/Cu joint interface. It can be found from Fig. 4.14a that the yield strength of the solder joint is about 24 MPa at the current strain rate, and the fracture strength is about 32 MPa. However, obvious stress relaxation has occurred even when the test was stopped at 20 MPa. Generally, the stress relaxation behavior is related with the applied stress and holding time. As shown in Fig. 4.14b, the relaxation rate is higher at higher stress, and both the stress and the relaxation rate decrease with increasing holding time. It has been reported that the stress relaxation behaviors of the Sn3.5Ag and Sn9Zn solder alloys are similar to the aforementioned behavior when the tensile tests were stopped at a constant stress [16], which suggests some validity for characterizing the dependence of tensile strength on time as an indicator of creep behavior. According to the shear strength and stress relaxation rates at different stresses, the stress ranges of the

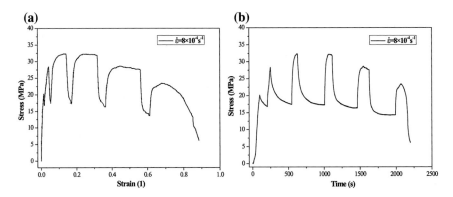

Fig. 4.14 Stress–strain and stress–time curves of Sn–Bi/Cu solder joints during the shear process. Reprinted from Ref. [34], Copyright 2011, with permission from Elsevier

creep-fatigue tests were chosen as 2–20 and 2–22 MPa to get proper lifetimes, and the holding time was chosen as 120 s since the stress relaxation occurs very fast. In that stress ranges, the Cu substrate only shows a very little elastic deformation, while the deformation and stress relaxation of the solder are obvious.

The macroscopic shear deformation and fracture behaviors of the Sn-58Bi/Cu solder joints are shown in Fig. 4.15, and the corresponding strains are tagged in each figure to indicate the deformation process. Figure 4.15a shows the morphology of the solder joint before the shear test, in which the surface of the SnBi solder is very flat. The strain in Fig. 4.15b is 0.112, and the solder joint has displayed a slight plastic deformation inside the solder, but not very obvious. In Fig. 4.15c, the strain is 0.211 and the plastic deformation becomes obvious in macroscale. With increasing strain, the plastic deformation inside the solder becomes more and more serious. At the strain of 0.273, a microscale slide crack has appeared at the corner of

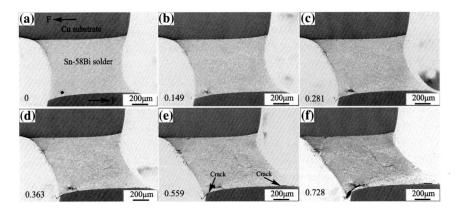

Fig. 4.15 Macroscopic shear deformation and fracture morphologies of Sn–Bi/Cu solder joints at different strains. Reprinted from Ref. [34], Copyright 2011, with permission from Elsevier

the solder joints, as shown in Fig. 4.15d, and obvious roughness at the surface of the solder can be observed. Figure 4.15e shows the deformation morphologies at the strain of 0.421, in which more cracks can be observed along the joint interface. Moreover, some microcracks inside the SnBi solder were also observed. In Fig. 4.15f, the slide crack at the interface has evolved into a macroscopic crack and the solder joint fractures along the joints interface shortly after that. During the shear fracture process, the strain concentrates inside the solder as the yield strength of the Cu substrate is very high, but there is no macroscopic cracking inside the solder even when its strain was up to 0.6, indicating that the SnBi solder has superior ductility in macroscale. However, as the strength and hardness of the two phases are quite different, serious deformation mismatch may easily occur between the two phases.

The microscopic deformation and fracture behaviors of the SnBi/Cu solder joints are shown in Fig. 4.16. Figure 4.16a–c shows the deformation behaviors of the SnBi solder during the shear process. In Fig. 4.16a the strain is 0.154, but serious plastic deformation has occurred inside the solder. The deformation mismatch between the Sn and Bi phases induces many steps at the surface, and some irregular microcracks were observed at the phase boundary. It can be predicated that these microcracks will decrease the resistance of the solder to plastic deformation. Figure 4.16b gives the deformation morphology of the solder at the strain of 0.307, which shows little difference from Fig. 4.16a, only a bit more serious. The deformation morphology at the strain of 0.660 is shown in Fig. 4.16c. Compared with that in Fig. 4.16a, the plastic deformation is much more serious, some solder even cracks into little grains. However, there is still no long cracking inside the solder. Because the microcracks inside the solder are irregular and have different orientations, they can hardly develop into long cracks. In fact, fracture of the solder joint often initiates at the Cu/solder joint interface. The deformation morphologies at the joint interface are shown in Fig. 4.16d–f. In Fig. 4.16d, the plastic deformation of

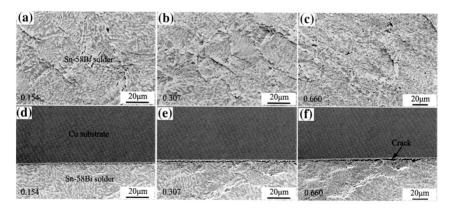

Fig. 4.16 a–c Deformation morphologies of Sn–Bi solder during the shear process; d–f deformation morphologies at the Sn–Bi/Cu joint interface. Reprinted from Ref. [34], Copyright 2011, with permission from Elsevier

the solder is not very serious, but a step appeared along the solder/Cu interface, because the transverse shrinkages of the solder and Cu are different. In Fig. 4.16e, the plastic deformation of the solder becomes obvious and the step has evolved into microcracks. With increasing strain, the microcracks become wider and connect to form lager cracks (see Fig. 4.16f), inducing a final fracture along the joint interface. According to the observations above, it is concluded that there is deformation mismatch between the two phases of SnBi solder, but the mismatch will not result in macroscale cracking. It is the deformation mismatch between the Cu substrate and solder that induces the final fracture of the solder joints along the joint interface.

4.4.2   Creep-Fatigue Fracture Behavior of Sn–Bi/Cu Solder Joints

The strain–cycle relationships of the solder joints during the creep-fatigue processes are shown in Fig. 4.17, and the two figures show the relationships in two different coordinates (see Fig. 4.17a, c). During the initial few cycles of the creep-fatigue

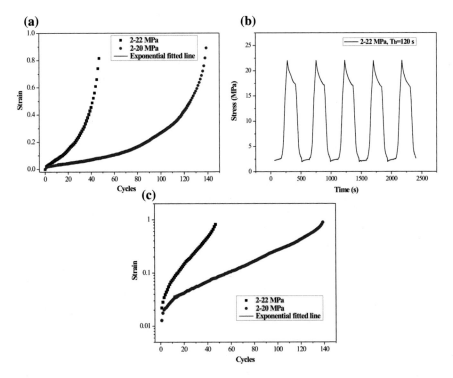

Fig. 4.17 Strain–cycle relationship of the creep-fatigue tests in **a** linear, **b** stress–time curves of a few cycles, and **c** exponential coordinate systems. Reprinted from Ref. [34], Copyright 2011, with permission from Elsevier

test, the strain increases rapidly but the increment per cycle decreases, as in Fig. 4.17a, which should result from the cyclic strain hardening of the solder during the fatigue process [17]. After a few cycles, the average strain increment per cycle decreases to a minimum value and then increases gradually with increasing cycles. As the strain is mainly contributed by plastic deformation of the solder, it is predicated that the capacity of the solder to resist the plastic deformation decreases with increasing cycles in that process. In microscale, it is the deformation mismatch inside the solder that bring in the damage and decrease the resistance of solder to plastic deformation. The stress–time curve of a few cycles in the stress range of 2–22 MPa is shown in Fig. 4.17b, it is found that there is obvious stress relaxation during the holding period at 22 MPa, and the relaxation rate becomes very low after holding for 120 s. Based on that, it can be found that the stress ranges and holding time have significant influence on the creep-fatigue behaviors. Since the strain–cycle curves are similar to the exponential lines, two exponential lines were drawn to fit the strain–cycle curves, and one can find that the fitted lines match well with the strain–cycle curves. Therefore, the strain is exhibited by exponential coordinate system in Fig. 4.17c. As in the figure, the strain–cycle shows a linear relationship during the intermediate stage. For the lattermost few cycles, the increase rate of strain becomes very high and deviates a little from the straight line, which should be due to the serious damage inside the solder joints and cracking at the joint interface prior to the final fracture. According to the results above, the process can be divided into three stages, i.e., strain hardening stage, exponential deforming stage, and final fracture stage, respectively. For the samples tested at different stress ranges, the fracture processes are similar, only the lifetime is different. According to the strain and cycle follows an exponential relationship, the exponential deforming stage can be described by an exponential function including the strain and cycles:

$$\varepsilon = A \, \exp(N/B) + C \tag{4.3}$$

where ε is the shear strain, N is the life cycle, and C is a constant. In the stress range of 2–20 MPa, A is 0.0123, B is 39.2, and C is 7.95×10^{-3}, while A is 0.0329, B is 18.3, and C is 1.16×10^{-2} in the stress range of 2–22 MPa. As the constant C is very little, the major difference is in A and B, which should be dominated by the stress range and holding time. Therefore, these two factors are the major extrinsic influencing factors on fatigue lives. As in Fig. 4.17, a little increase in the upper holding stress (20–22 MPa) can lead to a sharp decrease in the lifetime. If the upper holding stress increases a little more, the solder joints will fracture in very few cycles. In contrary, the lifetime will increase sharply when the stress decreases a little. For the solder joints in the electronic device, the thermal stress is induced by variation of the service temperature, and the upper holding stress is dominated by the maximum service temperature. Therefore, the creep-fatigue damage of the solder joints can be alleviated obviously by a slight change of the service temperature.

The macroscopic deformation and fracture morphologies of the solder joint tested in the stress range of 2–20 MPa are shown in Fig. 4.18; and the strain and

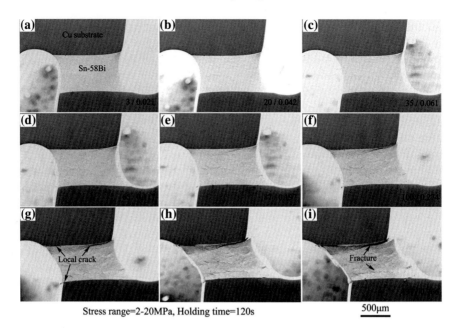

Stress range=2-20MPa, Holding time=120s   500μm

Fig. 4.18 Macroscopic creep-fatigue deformation and fracture morphologies of Sn–Bi/Cu solder joints. Reprinted from Ref. [34], Copyright 2011, with permission from Elsevier

cycles are also tagged in the images. The nine images are corresponding to the three stages of the creep-fatigue processes. In the first stage, there is no visible plastic deformation inside the solder in macroscale, as shown in Fig. 4.18a, although the strain hardening of the solder is obvious. Figure 4.18b–g shows the deformation morphologies of the second stage. In Fig. 4.18b–d, the plastic deformation of the solder can be observed, but not very serious, and the increase rate of strain is very low. In the latter process, the plastic deformation of the solder is much more serious and has become damaged rather than hardening (see Fig. 4.18e–g). The serious damage of the solder will decrease its resistance to plastic deformation, leading to a higher increase rate of strain. Besides, a deformation band appears at the corner of the solder joint in this stage and evolves into the primary cracking with increasing cycles. Prior to the final fracture, the cracking at the joint interface is serious and induces a sharp increase in strain, as in Fig. 4.18h, i, and final fracture occurs inside the solder near the joint interface.

It has been exhibited in Fig. 4.18 that the creep-fatigue fracture also occurs near the joint interface, which is similar to the shear fracture in macroscale. However, the crack initiation mechanisms are a little bit different in microscale. Figure 4.19 shows the interfacial deformation and the evolution process of the primary creep-fatigue cracks, and the corresponding strain of each figure is tagged at the corner. In Fig. 4.19a, b, the strain is very little, but a slight deformation band has appeared at the corner of the sample. As the strain concentration at the corner is

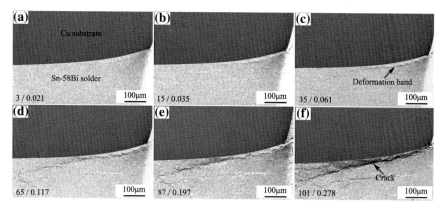

Fig. 4.19 Interfacial deformation and fracture behaviors of the Sn–Bi/Cu solder joint during the creep-fatigue process. Reprinted from Ref. [34], Copyright 2011, with permission from Elsevier

serious, the deformation band develops faster than that at the interior of the solder. As in Fig. 4.19c–e, the deformation band becomes longer and more serious with increasing strain. In Fig. 4.19f, the deformation band has evolved into a cracking near the joint interface, the damage rate of the solder joint accelerates in the later cycles and final fracture occurs along the crack. Therefore, it is the deformation bands near the joint interface and parallel to the shear direction that evolve into the primary creep-fatigue cracks. In contrast, the primary cracking of the shear fracture is evolved from the deformation mismatch between the solder and Cu substrate. Moreover, the deformation mechanisms of the Sn–Bi solder during the creep-fatigue processes are also a little bit different from the shear deformation process, which will be discussed in the next section.

4.4.3   Deformation Behavior of Sn–Bi Solder

As the strain of the solder joints is contributed by plastic deformation of the solder, the deformation mechanisms are necessary to be revealed in detail. The deformation morphologies of the Sn–58Bi solder at a certain region of the SnBi/Cu solder joint during the creep-fatigue process are shown in Fig. 4.20, and the strains are tagged in the images. Before the tests, the surface of the SnBi solder is very flat. After deformed for a few cycles, some roughness and streaks appear on the surface of the solder, as in Fig. 4.20a, b. With increasing cycles, the plastic deformation becomes more and more serious, and the roughness and deformation bands become obvious, as exhibited in Fig. 4.20c–d. Besides, it can be found that the streaks are actually deformation bands. It has been observed that the grain size of the air-cooled SnBi solder is about 100–200 μm [35], which is similar to the size of the particles encircled by the streaks. Through a careful examination on morphologies of the

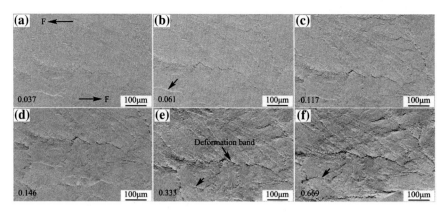

Fig. 4.20 Evolution of the deformation morphologies of Sn–Bi solder during the creep-fatigue process. Reprinted from Ref. [34], Copyright 2011, with permission from Elsevier

deformation bands, it is predicated that the deformation bands appear along the grain boundaries of the solder. In Fig. 4.20e, f, it is interesting to find that the deformation of the solder is not uniform. Around the deformation band, the plastic deformation is serious, while the deformation is little in the other regions. Moreover, if comparing the angle of the boundary indicated by the arrow, one can find a slight rotation of solder grain. To better understand the deformation mechanisms, the deformation morphologies around the solder grain are observed at higher magnification and shown in Fig. 4.21. The backscattering images were taken to reveal the deformation distribution, because only the microcracks in the serious plastic deformed areas can be identified as dark stripes in the backscattering images.

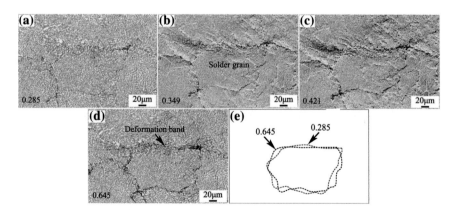

Fig. 4.21 Microscopic deformation morphologies of a Sn–Bi solder grain at different strains, **a** and **d** backscattering images; **e** configuration of the solder grain at different stains. Reprinted from Ref. [34], Copyright 2011, with permission from Elsevier

In Fig. 4.21a, a thin irregular circle of dark stripe is observed. Comparing it with the secondary electron images with similar strain (see Fig. 4.21b), it can be found that the dark cycle is a reflection of the serious plastic deformed area around the grain. In Fig. 4.21c, the strain is higher and the nonuniform of deformation distribution at different areas become more obvious. Besides, there is also a little deformation mismatch between the two phases inside the solder grain. However, that mismatch is far less serious and cannot be identified in the backscattering image. Figure 4.21d shows the backscattering electronic image of deformation at the strain of 0.403, in which the circle around the grain becomes wider, but there is no dark stripe inside the grain, indicating that the plastic deformation inside the grain is still little. Figure 4.21e shows the configuration of the grain in Fig. 4.21a, d stain, from which one can find the obvious deformation and rotation of the grain. Based on the observations in Figs. 4.20 and 4.21, it is predicated that the plastic deformation of the SnBi solder is concentrated along the grain boundary, therefore the solder deforms either in a grain rotation or in grain-boundary sliding mechanism.

There are two major creep deformation mechanisms in the Pb-bearing and the Pb-free solder alloys, i.e., dislocation climb and grain-boundary sliding, and their contribution to the creep strain depends on the strain rate, stress level, and homologous temperatures [30, 36, 37]. Generally, higher strain rate and lower homologous temperature correlate with greater dislocation climb, while a higher fraction of grain-boundary sliding can be expected at higher homologous temperature and lower strain rate [1]. As evidence, grain-boundary sliding has been observed in the Sn–37Pb solder during the thermal–mechanical cycling process [37]. Because of the high homologous temperatures and lower strain rate during the creep-fatigue process, the grain-boundary sliding should show a greater dominance in the creep deformation of the SnBi solders, which fits well with the phenomena shown in Figs. 4.20 and 4.21. As the creep deformation is mainly contributed by grain-boundary sliding, it is predicated that the grain size of the solder can affect the creep-fatigue resistance of the SnBi/Cu solder joints. Generally, grain-boundary sliding in creep process is easier to occur for metallic materials composed of fine grains [38–40]. Therefore, the creep resistance of the solder with larger grain size should be better. In fact, it has been reported that the increasing grain size of Sn–3Bi alloy leads to an improvement in the creep resistance in the temperature range of 303–333 K [41]. For the lead-free solder, an easy-operated method to coarsen the solder grain is to decrease the cooling rate during the reflowing process [42]. Moreover, a short-term thermal aging can also induce an obvious increase in grain size of the SnBi solder [43]. Through heat treating that can increase the grain size, the resistance of the SnBi solder to grain-boundary sliding and the creep-fatigue resistance of the SnBi/Cu solder joint can be enhanced.

4.5 Brief Summary

In this chapter, the creep-fatigue behavior of the Sn–4Ag/Cu and Sn–58Bi/Cu solder joints were tested through in situ observation, the creep-fatigue damage mechanisms were analyzed systematically, and major influencing factors on creep-fatigue life were discussed. The main conclusions are as follows:

1. Stress relaxation can easily occur during the shear process of the Sn–4Ag/Cu solder joint, when the stress is over 20 MPa. The creep-fatigue process consists of a strain hardening stage, a steady deformation stage, and an accelerating fracture stage. During the initial few cycles the strain increases rapidly, until strain hardening reaches a saturated state. After that the strain increases linearly with increasing number of cycles, deformation of the solder continues to develop, and strain concentration occurs around the solder/Cu_6Sn_5 interface and generates initial microcracks. When the microcracks evolve into long cracks creep-fatigue failure is accelerated and the solder joints fracture along the solder/Cu interface after a few more cycles. The strain on the solder joints is due to plastic deformation and creep of the solder. Grain subdivision occurs in the solder when the plastic strain reaches a certain threshold, then grain rotation and subdivision on a finer scale take place to accommodate further straining. Dislocation climb is predicated to be the major creep deformation mechanism, while grain-boundary sliding may also be promoted after the grain subdivision occurs.

2. The Sn–Bi solder in the Sn–Bi/Cu solder joint exhibits good ductility under shear loadings. Although there is serious deformation mismatch between the two phases of the Sn–Bi solder in microscale, no macroscale cracking can occur inside the solder. It is the deformation mismatch between the Cu substrate and solder that induces a step at the solder/IMC interface, which evolves into cracks with increasing strain and induces the shear fracture along the joint interface. The creep–fatigue fracture process can be divided into strain hardening stage, exponential deforming stage, and final fracture stage, and fracture occurs inside the solder near the joint interface. Since the ratios of the first and last stages are very low, its creep-fatigue process can be described by exponential function. The major creep deformation mechanism of the Sn–Bi solder is grain-boundary sliding. The plastic deformation concentrates at the grain boundary, while the deformation inside the solder grain is little.

3. Stress amplitude, average stress, and holding time are the major external factors influencing the creep-fatigue behavior. The microstructure of the solder can affect its deformation behavior, and the interfacial microstructure dominates crack initiation and final fracture. The size and dimensions of a solder joint can affect its creep-fatigue resistance by determining the strain concentration.

References

1. Zhang QK, Zhang ZF. In situ observations on creep fatigue fracture behavior of Sn–4Ag/Cu solder joints. Acta Mater. 2011;59:6017–28.
2. Evans JW. A guide to lead-free solders. 1st ed. London: Springer; 2005.
3. Ohguchi KI, Sasaki K, Ishibashi M. A quantitative evaluation of time-independent and time-dependent deformations of lead-free and lead-containing solder alloys. J Electron Mater. 2006;35:132–9.
4. Abtew M, Selvaduray G. Lead-free solders in microelectronics. Mater Sci Eng R. 2000;27:95–141.
5. Mathew MD, Yang H, Movva S, Murty KL. Creep deformation characteristics of tin and tin-based electronic solder alloys. Metall Mater Trans A. 2005;36:99–105.
6. Haung ML, Wang L, Wu CML. Creep behavior of eutectic Sn–Ag lead-free solder alloy. J Mater Res. 2002;17:2897–903.
7. Sharma P, Dasgupta A. Micro-mechanics of creep-fatigue damage in PB–SN solder due to thermal cycling—Part II: mechanistic insights and cyclic durability predictions from monotonic data. J Electron Pack. 2002;124:298–304.
8. Sharma P, Dasgupta A. Micro-mechanics of creep-fatigue damage in Pb–Sn solder due to thermal cycling-Part I: formulation. J Electron Pack. 2002;124:292–7.
9. Guo F, Choi S, Subramanian KN, Bieler TR, Lucas JP, Achari A, et al. Evaluation of creep behavior of near-eutectic Sn–Ag solders containing small amount of alloy additions. Mater Sci Eng A. 2003;351:190–9.
10. Zhang Q, Dasgupta A, Haswell P. Creep and high-temperature isothermal fatigue of Pb-free solders. Adv Electron Pack. 2003;1:955–60.
11. Kerr M, Chawla N. Creep deformation behavior of Sn–3.5Ag solder/Cu couple at small length scales. Acta Mater. 2004;52:4527–35.
12. Park S, Dhakal R, Lehman L, Cotts E. Measurement of deformations in SnAgCu solder interconnects under in situ thermal loading. Acta Mater. 2007;55:3253–60.
13. Sun Y, Liang J, Xu ZH, Wang GF, Li XD. In situ observation of small-scale deformation in a lead-free solder alloy. J Electron Mater. 2009;38:400–9.
14. Lang F, Tanaka H, Munegata O, Taguchi T, Narita T. The effect of strain rate and temperature on the tensile properties of Sn–3.5Ag solder. Mater Charact. 2005;54:223–9.
15. Shohji I, Yoshida T, Takahashi T, Hioki S. Tensile properties of Sn–Ag based lead-free solders and strain rate sensitivity. Mater Sci Eng A. 2004;366:50–5.
16. Takemoto T, Matsunawa A, Takahashi M. Tensile test for estimation of thermal fatigue properties of solder alloys. J Mater Sci. 1997;32:4077–84.
17. Kanchanomai C, Miyashita Y, Mutoh Y, Mannan SL. Influence of frequency on low cycle fatigue behavior of Pb-free solder 96.5Sn–3.5Ag. Mater Sci Eng A. 2003;345:90–98
18. Wild RN. Fatigue properties of solder joints. Weld J. 1972;51:521–6.
19. Solomon HD. Low cycle fatigue of Sn96 solder with reference to eutectic solder and a high Pb solder. J Electron Pack. 1991;113:102–8.
20. Deng X, Sidhu RS, Johnson P, Chawla N. Influence of reflow and thermal aging on the shear strength and fracture behavior of Sn–3.5Ag solder/Cu joints. Metall Mater Trans A. 2005;36A:55–64
21. Lee YH, Lee HT. Shear strength and interfacial microstructure of Sn–Ag–xNi/Cu single shear lap solder joints. Mater Sci Eng A. 2007;444:75–83.
22. Zhao J, Cheng CQ, Qi L, Chi CY. Kinetics of intermetallic compound layers and shear strength in Bi-bearing SnAgCu/Cu soldering couples. J Alloys Compd. 2009;473:382–8.
23. Zhang QK, Zhang ZF. Fracture mechanism and strength-influencing factors of Cu/Sn–4Ag solder joints aged for different times. J Alloy Compd. 2009;485:853–61.
24. Zhang QK, Zou HF, Zhang ZF. Tensile and fatigue behaviors of aged Cu/Sn–4Ag solder joints. J Electron Mater. 2009;38:852–9.

25. Wu X, Tao N, Hong Y, Xu B, Lu J, Lu K. Microstructure and evolution of mechanically-induced ultra fine grain in surface layer of AL-alloy subjected to USSP. Acta Mater. 2002;50:2075–84.
26. Liu Q, Jensen DJ, Hansen N. Effect of grain orientation on deformation structure in cold-rolled polycrystalline aluminum. Acta Mater. 1998;46:5819–38.
27. Hansen N, Huang X, Hughes DA. Microstructural evolution and hardening parameters. Mater Sci Eng A. 2001;317:3–11.
28. Kashyap BP, Murty GS. Experimental constitutive relations for high-temperature deformation of a Pb–Sn eutectic alloy. Mater Sci Eng. 1981;50:205–13.
29. Shine MC, Fox LR. Fatigue of solder joints in surface mount devices, STP 942. Philadelphia, PA: ASTM; 1988. p. 588–610.
30. Mavoori H, Chin J, Vayman S, Moran B, Keer L, Fine M. Creep, stress relaxation, and plastic deformation in Sn–Ag and Sn–Zn eutectic solders. J Electron Mater. 1997;26:783–90.
31. Yeung B, Jang JW. Correlation between mechanical tensile properties and microstructure of eutectic Sn–3.5Ag solder. J Mater Sci Lett. 2002;21:723–6.
32. Dao M, Chollacoop N, Van Vliet KJ, Venkatesh TA, Suresh S. Computational modeling of the forward and reverse problems in instrumented sharp indentation. Acta Mater. 2001;49:3899–918.
33. Deng X, Chawla N, Chawla KK, Koopman M. Deformation behavior of (Cu, Ag)–Sn intermetallics by nanoindentation. Acta Mater. 2004;52:4291–303.
34. Matin MA, Vellinga WP, Geers MGD. Microstructure evolution in a Pb-free solder alloy during mechanical fatigue. Mater Sci Eng A. 2006;431:166–74.
35. Mahmudi R, Geranmayeh AR, Mahmoodi SR, Khalatbari A. Room-temperature indentation creep of lead-free Sn–Bi solder alloys. J Mater Sci Mater Electron. 2007;18:1071–8.
36. Dutta I. A constitutive model for creep of lead-free solders undergoing strain-enhanced microstructural coarsening: a first report. J Electron Mater. 2003;33:201–7
37. Telang AU, Bieler TR, Crimp MA. Grain boundary sliding on near-7 degrees, 14 degrees, and 22 degrees special boundaries during thermornechanical cycling in surface-mount lead-free solder joint specimens. Mater Sci Eng A. 2006;421:22–34.
38. Sherby OD, Taleff EM. Influence of grain size, solute atoms and second-phase particles on creep behavior of polycrystalline solids. Mater Sci Eng A. 2002;322:89–99.
39. Nabarro FRN. Creep in commercially pure metals. Acta Mater. 2006;54:263–95.
40. Padmanabhan KA. Grain boundary sliding controlled flow and its relevance to superplasticity in metals, alloys, ceramics and intermetallics and strain-rate dependent flow in nanostructured materials. J Mater Sci. 2009;44:2226–38.
41. Abd El-Rehim AF, Effect of grain size on the primary and secondary creep behavior of Sn–3 wt% Bi alloy. J Mater Sci. 2008;43:1444–50
42. Kim KS, Huh SH, Suganuma K. Effects of cooling speed on microstructure and tensile properties of Sn–Ag–Cu alloys. Mater Sci Eng A. 2002;333:106–14.
43. Zhang QK, Zhu QS, Zou HF, Zhang ZF. Fatigue fracture mechanisms of Cu/lead-free solders interfaces. Mater Sci Eng A. 2010;527:1367–76.

Chapter 5
Thermal Fatigue Behavior of Sn–Ag/Cu Solder Joints

5.1 Introduction

In the electronic components, the primary strain subjected by the solder joints is resulted from the difference in the CTE among the chip, chip-carrier, and the circuit board. As shown in Fig. 5.1, when the component is heated from the equilibrium temperature, the solder joints will suffer a shear strain, which reversed when the temperature decreases [1]. Since the electronic equipments are periodically turned on and off, cyclic strain occurs inside the solder joints, leading to the thermal fatigue damage [2]. As a result, understandings on thermal fatigue damage mechanisms of the solder joints are very important for evaluating their mechanical reliability. During the thermal fatigue process, common plastic deformation is the major deformation mechanism of the solder. Besides, since the homologous temperature (T/T_m) of the Sn-based solders used in electronic assembly usually exceeds 0.6 even at room temperature, both plastic deformation and creep can occur inside the solder during the thermal cycling [3]. The microstructure evolutions such as recovery, grain subdivision, and recrystallization have also been observed in the deformed solder [4–9]. Therefore, thermal fatigue damage of the solder joints is a complex process involving plastic deformation, creep, microstructure evolutions, and fracture.

Thus far, many related studies have focused on thermal fatigue behaviors of the solder joints, the flip-chip or single-lap solder joints are usually used as the test samples, and the general thermal fatigue damage mechanisms and the influencing factors have been proposed [5, 10–16]. It was found that cracks usually initiate and propagate at the tip of the lap-shear solder joints; geometry of shear joint was significant in determining the crack initiation [14, 17]. The microcracks usually initiate around the free surface of the joint interface, and then propagate along the joint interface or along the boundaries of the β-Sn grains. However, the strain amplitudes used in most of these studies are too high, making the thermal fatigue damage quite different from that in the real service environment. Besides, the dynamic thermal fatigue damage process has not been well revealed.

© Springer-Verlag Berlin Heidelberg 2016
Q. Zhang, *Investigations on Microstructure and Mechanical Properties of the Cu/Pb-free Solder Joint Interfaces*, Springer Theses,
DOI 10.1007/978-3-662-48823-2_5

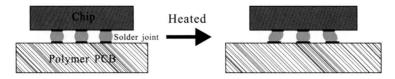

Fig. 5.1 Thermal deformation of solder joints in electronic component

For the reasons above, in this study, thermal fatigue behaviors of the Pb-free solder joints at low strain amplitude were comprehensively investigated. The thermal strains were applied on the solder joints using special-designed clamps and a temperature-controlled cabinet. To better simulate the service condition of the solder joint in electronic device, the samples were tested at low strain amplitudes. Through revealing the damage morphologies of the solder joints deformed for different cycles and the evolutions in microstructures of the solder, visualized failure processes were obtained. The joint interfaces of the specimens are relatively small (1 mm × 1 mm × 0.5 mm), in order to make them more similar to the solder joints in the microelectronic devices. Based on the observation results on deformation morphologies and microstructure evolution of the solder, understandings on thermal fatigue mechanisms of the solder joints at low strain amplitude are provided, and the influencing factors on thermal fatigue resistance are discussed.

5.2 Experimental Procedure

The Sn–4Ag/Cu solder joints were employed as example in this study. The cold-drawn oxygen-free-high conductivity Cu with a yielded strength of about 300 MPa was chosen as the substrate material. The solder alloy was prepared by smelting high-purity (>99.99 %) tin, copper, and silver at 800 °C for 30 min in vacuum. The test samples are the same to that used in Chap. 4, the preparation process is shown in Fig. 5.2 [18]. The Cu substrate was firstly spark cut into small blocks (15 mm × 10 mm × 4 mm) with a step at one end, and then the surfaces at the steps were ground and electrolytically polished. After air drying, a flux was dispersed on the polished area to enhance wetting and minimize oxide formation. The steps of two Cu blocks were butt to butt, a Sn–4Ag alloy sheet was sandwiched between them and two graphite plates were clamped on their sides to avoid the outflow of the molten solder. The prepared samples were put in an oven with a temperature of 260 °C, kept for 8 min after the melting of the solder and then cooled down in air. Then, the solder joints were sliced by spark cutting, and their side surfaces were grounded and carefully polished for interfacial observations.

The clamp designed for the thermal fatigue test is shown in Fig. 5.3. As in figure, the solder joint is sandwiched between two polyvinyl chloride (PVC) plates, and its two ends are padded with sand paper and fastened with bolt and nut to avoid slip. Since the CTEs of Cu and PVC are different, their thermal deformations are

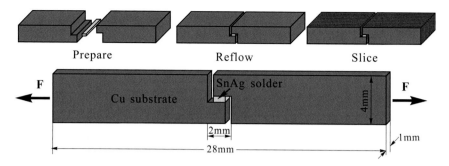

Fig. 5.2 Preparation process and dimension of testing specimens. Reprinted from Ref. [18], Copyright 2013, with permission from Elsevier

Fig. 5.3 Clamp of thermal fatigue test. Reprinted from Ref. [18], Copyright 2013, with permission from Elsevier

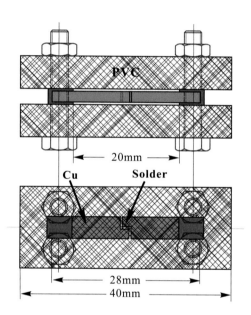

also quite different when the temperature changes and the solder joints will be driven to deform. Besides, since the Cu substrate has much higher yield strength than the Sn–Ag solder, it only exhibits very slight elastic deformation when the solder joint deforms, and the major deformation of the solder joint concentrates inside the solder. The average thermal strain suffered by the solder joint can be calculated by the follow equation:

$$\gamma = (\alpha_{th1} - \alpha_{th2})l\varDelta T/t \tag{5.1}$$

where α_{th1} is CTE of PVC, α_{th2} is CTE of the Cu substrate, $\alpha_{th1} = 8\times10^{-5}\ °C^{-1}$, and $\alpha_{th2} = 1.7 \times 10^{-5}\ °C^{-1}$. $\varDelta T$ is the temperature amplitude, l is the distance between the two ends of the solder joint (20 mm), and t is the thickness of the solder in the

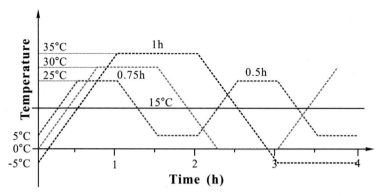

Fig. 5.4 Temperature cycling curves of thermal fatigue tests. Reprinted from Ref. [18], Copyright 2013, with permission from Elsevier

solder joint (0.5 mm). Three temperature ranges, i.e., 5–25 °C, 0–30 °C, −5 to 35 °C, were chosen to make a comparison, as in Fig. 5.4. The solder joints were clamped at a temperature of about 15 °C, and the strain amplitudes are calculated to be ±2.5, ±3.75, and ± 5 % at three temperature amplitudes, and the periods are 2, 3, and 4 h, respectively. It should be noticed that the "real strain" is actually the average nominal strain, because the deformation in the solder joint is not uniform. Since the thermal strain rate depends linearly on the temperature changing rate, it can be calculated that the strain rates at both the three temperature amplitudes are $2.78 \times 10^{-5} \text{ s}^{-1}$.

The thermal fatigue tests were conducted in a BE-TH-80M3 temperature-controlled cabinet; two samples were tested at the each condition. After deformed for a certain cycles, the nuts were loosen, the solder joints were taken down form the clamp, and their deformation morphologies were observed by ZEISS Supra 35 field emission SEM. Since there is always strain concentration around the joint interface [1, 19, 20], the interfacial deformation morphologies were observed with emphases. To reveal the deformation mechanisms of the solder during the thermal fatigue process, microstructures of the solder were characterized by EBSD (HKL Channel software version 5). Fracture surfaces of the solder joints were also observed after the tests.

5.3 Thermal Fatigue Behavior

5.3.1 Thermal Fatigue Behavior at Low Strain Amplitude

Figure 5.5 shows the macroscopic deformation morphologies of the solder joint cycled at 5–25 °C for different times, with the cycles marked in each figure. Before the thermal fatigue test, surface of the sample is flat and smooth, no deformation or

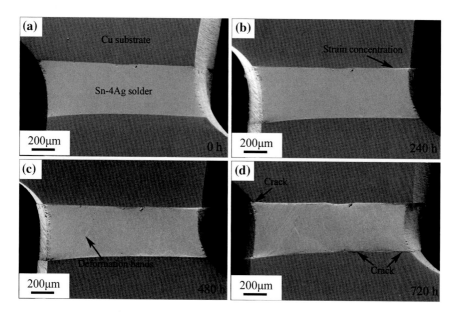

Fig. 5.5 Macroscopic deformation morphologies of Sn–4Ag/Cu solder joint cycled at 5–25 °C: **a** 0 cycle, **b** 120 cycles, **c** 240 cycles, and **d** 360 cycles. Reprinted from Ref. [18], Copyright 2013, with permission from Elsevier

damage was observed, as in Fig. 5.5a. After deformed for 240 h, slight deformation bands appeared in the solder (see Fig. 5.5b), and obvious strain concentration occurred around the interface. Figure 5.5c shows the deformation morphologies after cycled for 480 h, in which the deformation bands have became very obvious, and the interfacial strain concentration is more serious. After 720 h, deformation of the solder increases a bit, and it can be found that deformation at different areas is not uniform, as in Fig. 5.5d, the bands show different orientations, which should be attributed to the difference in orientations of the solder grains. Besides, some microcracks have appeared at the joint interface, and slide cracking occurs at the corner of the solder joint, because the strain concentration occurs around the joint interface [21, 22], especially at the corner of the solder joints [14, 17]. Since the strain amplitude suffered by the solder joint cycled at 5–25 °C is very low, the thermal fatigue damage develops slowly. For the same reason, damage behavior of the sample should be more similar to that of the solder joints in the real service environment. To better reveal, the thermal fatigue behaviors, morphologies of the solder, and the joint interface cycled for different times were observed at higher magnification, and microstructure of the solder was characterized by EBSD.

The interfacial deformation morphologies at the corner of the solder joints cycled at 5–25 °C are shown in Fig. 5.6. As in Fig. 5.6a, no deformation or damage can be observed before the test even at higher magnification. After deformed for 240 h, a group of parallel deformation bands appear at the surface of the solder, and obvious deformation mismatched occur around the upper right joint interface, as in

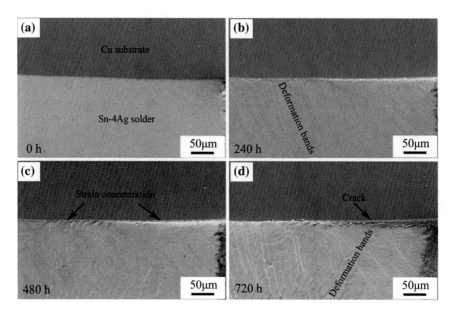

Fig. 5.6 Interfacial deformation morphologies of Sn–4Ag/Cu solder joint cycled at 5–25 °C: **a** 0 cycle, **b** 120 cycles, **c** 240 cycles, and **d** 360 cycles. Reprinted from Ref. [18], Copyright 2013, with permission from Elsevier

Fig. 5.6b. The deformation bands are generally recognized to be slip bands formed through dislocation slip and multiplication [14, 23]. Besides, the deformation morphologies at the left and right sides of figure are quite different, induced by the difference in grain orientations. With increasing cycles, the deformation keeps developing, and a new group of wavelike deformation bands appeared at the lower region after 480 h (see Fig. 5.6c). The new deformation bands are not so regular as the former group and are usually recognized as shear bands formed by relative motion between different parts of the solder grain [1, 15]. After 720 h, the deformation bands are much serious, and the solder around the corner has fractured, forming some microcracks, as in Fig. 5.6d.

Figure 5.7 shows the deformation morphologies of a selected area in the solder joint cycled at 5–25 °C. Before the thermal cycling, morphologies of the solder at different areas show little difference, as in Fig. 5.7a. After cycled for 240 h, two groups of slip bands intersecting an angle appear, each group consists of some parallel bands (see Fig. 5.7b). Since the orientations of the two groups of slip bands are quite different, they should correspond to the solder grains with obvious misorientation, and the interface is the grain boundary. Figure 5.7c shows the solder deformed for 480 h, in which the slip bands become wider and thicker, but no new slip bands appear. After 720 h, the slip bands are obvious, as in Fig. 5.7d, and some slight wavelike undulations appear in the solder, which might be formed due to relative motion between the thin grains or sub-grains. In general, it is concluded that different deformation mechanisms occur inside the solder during the thermal fatigue

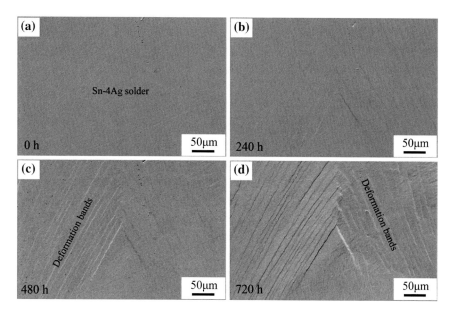

Fig. 5.7 Deformation morphologies of solder in Sn–4Ag/Cu solder joint cycled at 5–25 °C: **a** 0 cycle, **b** 120 cycles, **c** 240 cycles, and **d** 360 cycles. Reprinted from Ref. [18], Copyright 2013, with permission from Elsevier

process, forming different surface morphologies. Besides, deformation of the solder in this study is generally far less serious than that in the thermal fatigue tests reported before, and no fracture occurs inside the solder. Similar to the thermal fatigue at high strain amplitude [15, 23], deformation of the solder has only minimal effect on the thermal fatigue damage cracking at the current strain amplitude.

Since the melting points of the Sn-based solders are very low, recovery and recrystallization can easily occur in them. The microstructure evolution behaviors depend on the deformation history of the solder and can affect its further deformation [24]. Therefore, it is necessary to investigate the evolution in microstructure of the solder during the thermal fatigue process. The grain maps of the solder cycled at 5–25 °C for different times are shown in Fig. 5.8, in which different colors correspond to different grain orientations. The low-angle grain boundaries ($2° < \Delta\theta < 15°$) are indicated with white lines and high-angle boundaries ($\Delta\theta > 15°$) with black lines. In the characterized region, there are 4 large grains separated by high-angle grain boundaries before the test, each of them consists of thin grains separated by low-angle grain boundaries, as shown in Fig. 5.8a, the large grains are numbered (I–IV) to distinguish and describe them. The grain maps of the solder deformed for 240 h are shown in Fig. 5.8b. As in figure, grains III and IV change little, while grain II on the left became much thinner, because it was absorbed by grain I. After 480 h, the contrast in colors of the thin grains is more obvious, i.e., the misorientation angles between them increase, which should be induced by relative rotation between them (see Fig. 5.8c). Figure 5.8d shows the maps after 720 h, it

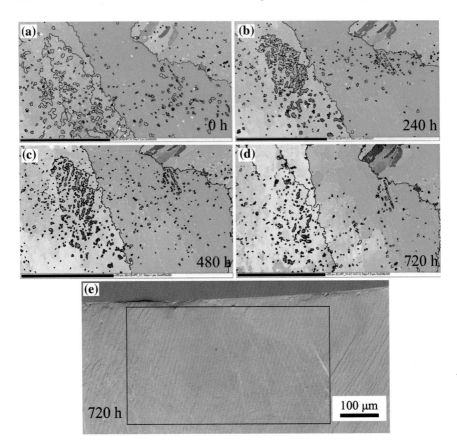

Fig. 5.8 Evolution in microstructure of solder in Sn–4Ag/Cu solder joint cycled at 5–25 °C: **a** 0 cycle, **b** 120 cycles, **c** 240 cycles, and **d, e** 360 cycles. Reprinted from Ref. [18], Copyright 2013, with permission from Elsevier

can be found that the misorientations between the thin grains keep increase, and the orientations of the large grains also change obviously compared with their initial state, because grain rotation occurs during the deforming process. The grain coalesce has also been observed during the creep-fatigue process of the Sn–Ag/Cu solder joints, which accomplished in a short time, while it took much longer time when the solder joint was cycled at 5–25 °C. Since the strain energy is the major driving force of grain migration, the low migration rate indicates that the stain energy in the solder is very low. For the grains shows no grain migration, the high-angle grain boundaries are stable in general, but their shape and orientation still change a little, because there is relative sliding between these boundaries. The area enclosed by a rectangle in Fig. 5.8e corresponds to the grain map in Fig. 5.8d. It is obvious that the orientations of the deformation bands in different large grains are quite different, while that of the bands in the same large grain are the same, indicating that the slip bands can easily get across the low-angle grain boundaries, but not the high-angle grain boundaries.

5.3.2   Thermal Fatigue Behavior at Medium Strain Amplitude

The macroscopic deformation morphologies of the solder joint cycled at 0–30 °C for different times are shown in Fig. 5.9, with the cycles marked in each figure. Figure 5.9a shows the morphology before the test. After cycled for 120 h, the solder showed slight deformation, and obvious interfacial strain concentration occurs at the upper left corner of the solder joint (see Fig. 5.9b). After 240 h, deformation of the solder increases obviously, and sliding cracks have appeared, as in Fig. 5.9c. Figure 5.9d shows the deformation morphology after 300 h, in which serious fracture occurred along the upper joint interface. Due to the higher temperature/strain amplitude, damage rate of the solder joint cycled at 0–30 °C is much higher than that cycled at 5–25 °C, the joint interface fractured after 300 h, and deformation of the solder is also more serious.

The interfacial deformation morphologies of the solder joint cycled at 0–30 °C are shown in Fig. 5.10. Before the thermal fatigue test, there is no damage around the joint interface, as in Fig. 5.10a. After cycled for 120 h, obvious deformation

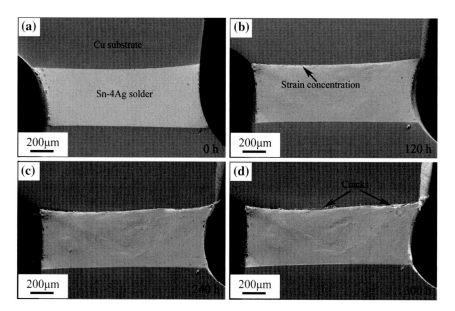

Fig. 5.9 Macroscopic deformation morphologies of Sn–4Ag/Cu solder joint cycled at 0–30 °C: **a** 0 cycle, **b** 40 cycles, **c** 80 cycles, and **d** 100 cycles. Reprinted from Ref. [18], Copyright 2013, with permission from Elsevier

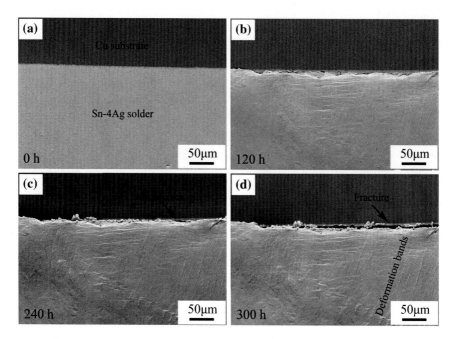

Fig. 5.10 Interfacial deformation morphologies of Sn–4Ag/Cu solder joint cycled at 0–30 °C: **a** 0 cycle, **b** 40 cycles, **c** 80 cycles, and **d** 100 cycles. Reprinted from Ref. [18], Copyright 2013, with permission from Elsevier

bands appeared in the solder (see Fig. 5.10b), and the strain concentration and deformation mismatch initiate some damage at the solder/IMC interface. After 240 h, the interfacial damage evolved into sliding crack, as in Fig. 5.10c. Deformation of the solder around the joint interface has broke up and been extruded. Besides, there is an obvious boundary in the solder; deformation bands can be found on the right side, while there is only slight surface undulation on the left side. Therefore, the deformation boundary should be a high-angle grain boundary, the grains at the two sides deform in different mechanisms, resulting in different surface morphologies. Figure 5.10d shows the morphology after 300 h, in which the sliding crack has evolved into serious fracture, while deformation of the solder shows little increase. Since the deformation usually tends to concentrate around the crack tips, the interfacial damage increases sharply, while deformation of the interior solder changes little. In general, deformation of solder and interfacial strain concentration are much serious for the solder joint cycled at 0–30 °C; crush and extrusion occur inside the solder around the joint interface.

The grain maps of the solder cycled at 0–30 °C for different times are shown in Fig. 5.11. Similar to that in Fig. 5.8, grain-boundary migration and grain coalesce were also observed in the solder, but the grain boundaries become stable in a relatively short time (see Fig. 5.11a–c), which should attributes to the higher driving force, i.e., the higher strain energy in the solder. Based on that, it is predicated that

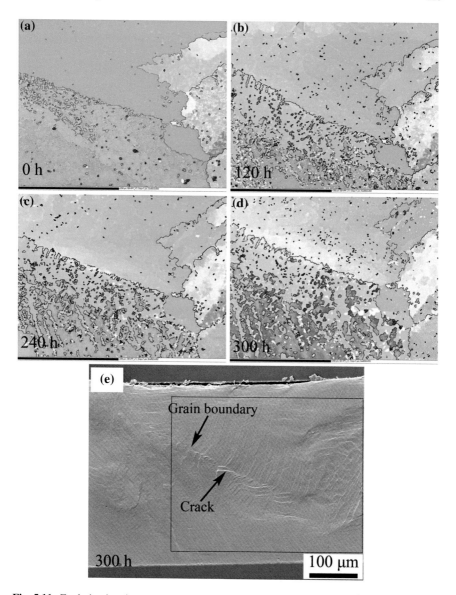

Fig. 5.11 Evolution in microstructure of solder in Sn–4Ag/Cu solder joint cycled at 0–30 °C: **a** 0 cycle, **b** 40 cycles, **c** 80 cycles, and **d**, **e** 100 cycles. Reprinted from Ref. [18], Copyright 2013, with permission from Elsevier

plastic deformation of the solder is higher at this strain amplitude. Since the interfacial strain concentration at the crack tip is serious, deformation degree of the solder shows little increase in the latter stage of the thermal fatigue process, thus evolution in microstructure of the solder is also not obvious. Comparing Fig. 5.11d, e, it is found that the slip bands pile up at the grain boundary, making the deformation around the grain boundary very serious, and forming the microcracks. The orientations and deformation degree of the bands in different grains are quite different.

5.3.3 Thermal Fatigue Behavior at High Strain Amplitude

Microscopic damage processes of the solder joint cycled at −5 to 35 °C are shown in Fig. 5.12. As in Fig. 5.12a, there is no damage or defect in the solder joint before the test. After 60 h, however, serious deformation has appeared. There is also difference in deformation morphologies at different regions of the solder and strain concentration around the joint interface (see Fig. 5.12b). After 120 h, deformation degree of solder increases a little bit, the boundaries become more obvious, and serious fracture along the interface occurs, as in Fig. 5.12c. Due to the high strain amplitude, damage of the solder joint cycled at −5 to 35 °C develops very fast, and the interfacial strain concentration is also more serious. After 180 h, the solder joint has fractured along the joint interface, while deformation of the solder changes little, as in Fig. 5.12d. Through comparing the damage behaviors of the solder joints cycled at the three strain amplitudes, it can be concluded that the fatigue life is very sensitive to the strain amplitude. The strain amplitude of the solder joint cycled at −5 to 35 °C is twice as high as the solder joint cycled at 5–25 °C, while damage of

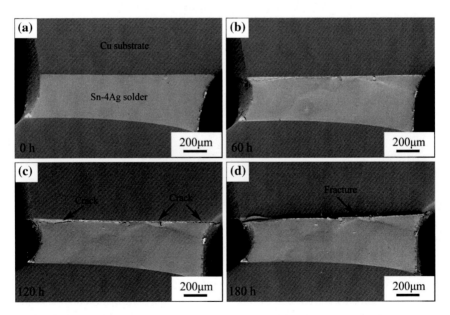

Fig. 5.12 Macroscopic deformation morphologies of Sn–4Ag/Cu solder joint cycled at −5 to 35 ° C: **a** 0 cycle, **b** 15 cycles, **c** 30 cycles, and **d** 45 cycles. Reprinted from Ref. [18], Copyright 2013, with permission from Elsevier

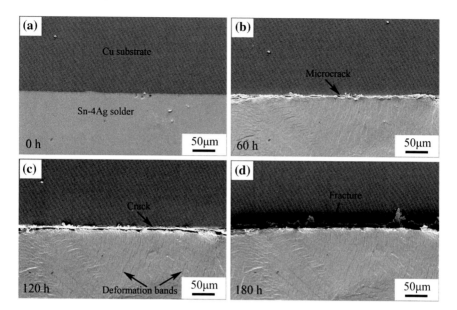

Fig. 5.13 Interfacial deformation morphologies of Sn–4Ag/Cu solder joint cycled at −5 to 35 °C: **a** 0 cycle, **b** 15 cycles, **c** 30 cycles, and **d** 45 cycles. Reprinted from Ref. [18], Copyright 2013, with permission from Elsevier

the former solder joint after 180 h/45 cycles is more serious than that of the latter deformed for 720 h/360 cycles, indicating that fatigue life of the former should be less than a quarter of the latter.

The interfacial deformation morphologies of the solder joint cycled at −5 to 35 °C are shown in Fig. 5.13. Figure 5.13a shows the morphology before the test. After 60 h, serious deformation has appear inside the solder, and a few groups of deformation bands emerged, which are deeper than that shown ahead (see Fig. 5.13b). As the deformation and strain concentration are much serious in this condition, microcracks initiated after only 60 cycles. The interfacial cracks were wider and a little solder was extruded after 120 h, as in Fig. 5.13c. After cycled for 180 h, the joint interface has fractured (see Fig. 5.13d). Due to the friction between the two sides of the crack, some solder was extruded. Compared with the deformation behaviors shown in Figs. 5.6 and 5.10, the interfacial damage rate is much higher, the interfacial deformation is more serious and a lot of solder is extruded, while the general damage processes are similar.

5.3.4 Thermal Fatigue Fracture Surface

To comprehensively reveal the thermal fatigue damage behaviors, the fracture surfaces were also observed. Figure 5.14 shows the fracture morphologies of a

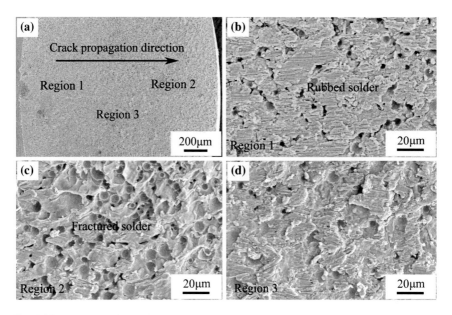

Fig. 5.14 Fracture surface of the solder joints: **a** macroscopic fracture surfaces, microscopic fracture surfaces of **b** crack propagation region, **c** accelerating fracture region, and **d** transition region. Reprinted from Ref. [18], Copyright 2013, with permission from Elsevier

solder joint cycled at 0–30 °C. The macroscopic fracture surface is shown in Fig. 5.14a. As in figure, there is a thin layer of solder covering the fracture surface, and the difference in morphologies of the fractured solder can be noticed even at low magnification. The fracture surface consist of two regions, the surface on the right side is a bit rougher. At high magnification, differences between the two regions are more obvious, as exhibited in Fig. 5.14b, c. On the left side (region 1), the trace of friction can be found (see Fig. 5.14b). Therefore, this region should be formed at the earlier stage, and the friction trace was formed when the two sides of the crack rubbed with each other during the latter cycles, making the traces parallel with the loading direction. Since region 1 ranges from the left edge of the fracture surface, it can be confirmed that the cracks initiate at the corner of the solder joint. In contrast, the trace of friction is not obvious in region 2, while some tearing ridges were observed, as in Fig. 5.14c. Since the crack propagation rate increases sharply at the latter stage of the thermal fatigue process, final fracture occurs shortly after the crack propagated to region 2, thus the fracture surface underwent little friction, making it similar to the shear fracture surface [25, 26]. Besides, small voids can be found on the fracture surface. During the thermal fatigue process, there should be relative motion between the solder and the interfacial IMC layer, which gradually initiates some microcracks and inducing small voids on the fracture surface. Figure 5.14d shows the morphology of the transition area between region 1 and region 2, with both friction trace and tearing ridges observed in it. In addition, there is no cracked Cu–Sn intermetallic compound at the fracture surface, indicating that the IMC layer did not fracture during the thermal fatigue process, because the thermal stress and strain rate are very low, while fracture of the IMC layer only occurs at high stress and strain rate [26, 27].

5.4 Thermal Fatigue Damage Mechanisms

5.4.1 Thermal Fatigue Damage Process

Based on the observation results exhibited above, the thermal fatigue process of the Sn–4Ag/Cu solder joints can be divided into two stages according to their deformation and fracture behaviors, i.e., the strain hardening stage and the accelerating fracture stage. Figure 5.15 shows a qualitative illustration on the damage process, in which the abscissa indicates the thermal cycles, and the ordinate represents the nominal strength of the solder joint. The first stage is the strain hardening stage. In this stage, strain hardening of the solder keeps developing with increasing cycles, making the strength of the solder increase gradually. On the other hand, the interfacial damage is still slight, and no cracks appear at the interface, making the nominal strength of the solder joint increases in this stage. It should be noticed that dynamic recovery and stress relaxation occur rapidly in the Sn-based solder even at

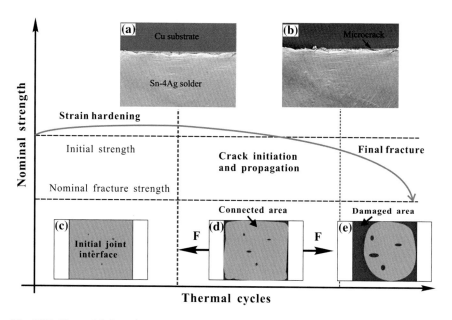

Fig. 5.15 Thermal fatigue damage mechanisms of the Cu/lead-free solder joints. Reprinted from Ref. [18], Copyright 2013, with permission from Elsevier

room temperature [24, 28–30], therefore strain hardening actually occurs through the whole thermal fatigue process, and it can achieve a dynamic equilibrium state with recovery [1, 28].

Due to the stain concentration, deformation of the solder around the joint interface is far more serious than average [1, 19, 20], especially at the corner, and can not be eliminated through recovery. When it increases to a certain degree, the accumulated damage will generate some microcracks at the joint interface and decrease the loading capacity of the solder joint. As a symptom, it signifies that the thermal fatigue enters the accelerating fracture stage (see Fig. 5.15a). In this stage, the damage rate increases gradually, because the real stress subjected by the solder joint increases with the crack propagation. Therefore, the nominal strength of the solder joint shows an accelerating decrease tendency. During the crack propagation process, friction between the two sides of the cracks forms the trace at the fracture surfaces, and some solder close to the interface was extruded, as in Fig. 5.15b. When the microcracks evolve into long crack, the damage rate increases sharply. Morphology of the final fracture region is similar to the shear fracture morphologies. The damage process of the joint interface is shown in Fig. 5.15c–e, in which the red area is the damaged area. Theoretically, the final fracture condition can be expressed as:

$$\tau_r = \tau_{max}/(1 - \Delta s) = \tau_f \qquad (5.2)$$

where τ_{max} is the peak stress, τ_r is the real shear stress, τ_f is the shear fracture strength of the solder joint, and Δs is a parameter describing the decreasing ratio in area of the joint interface. However, the solder joint can hardly fracture thoroughly, because the solder has superior ductility and only fracture at very high strain [1].

5.4.2 Deformation Mechanisms of Solder

During the thermal fatigue process, deformation mechanisms of the solder are complex, but are generally regarded to be a sum of plastic deformation and creep [5, 8–10, 13–16, 31]. To comprehensively reveal the thermal fatigue damage mechanisms of the solder joint, it is necessary to discuss deformation behaviors of the solder, thus the thermal deformation curves of the solder joint were proposed firstly. Although there is no strain gage used in this study, deformation behavior of the solder joint can be estimated through analyzing the deformation behavior of similar solder joints. It has been well accepted that shear strength of the solder joint and the strain rate follow the following relationship [32–34]:

$$\tau = C\dot{\gamma}^m \qquad (5.3)$$

where τ is the shear strength, $\dot{\gamma}$ is the shear strain rate, m is a sensitive factor related to the solder, and C is a constant. For the Sn–Ag solder, m is around 0.08 [32].

As the shear strength of the Sn–4Ag/Cu solder joint is 32 MPa at the strain rate of 2.5×10^{-3} s^{-1} [1], and the strain rate during the thermal fatigue process is 2.78×10^{-5} s^{-1}, shear strength of the solder joint is calculated to be 22.3 MPa in this study. It has been reported that strength of the solder becomes constant when the strain increases to a certain value, i.e., the strain hardening becomes saturated at a certain strain, and the saturation strain is related to the strain rate [28]. According to the saturation strain of the Sn–4Ag/Cu solder joint at a strain rate of 2.5×10^{-3} s^{-1}, and that of the Sn–Ag–Cu/Cu solder joint at a strain rate of 1.0×10^{-4} s^{-1} [1, 28], the saturation strain of the Sn–4Ag/Cu solder joint is predicated to be around 2–3 % in this study. Furthermore, based on the stress relaxation behavior of the Sn–4Ag/Cu solder joint at 25 °C [30], it is estimated that the flow stress of the Sn–4Ag/Cu solder joint will decrease to about 15 MPa after dwelling at 25–35 °C for 0.5–1 h.

According to the discussions above, the "nominal stress"-"real strain" curve of the Sn–4Ag/Cu solder joint at the early stage of the thermal fatigue process is estimated and shown in Fig. 5.16a. For the solder joint cycled at 0–30 °C, the strain amplitude is ±3.75 %. At the initial thermal fatigue cycle, the nominal stress/strain equal to the real stress/strain. When the strain exceeds 2 %, the strain hardening becomes saturated, and the saturated stress is around 22.3 MPa. Then, after dwell at 30 °C for 45 min, the stress decreased to 15 MPa. Then, deformation of the solder joint reversed when the temperature decreases to 0 °C. However, since there is residual strain hardening in the solder, the nominal saturation stress will increase within the initial thermal cycles, while the hardening strain decreases a little. After a certain cycles, the strain hardening and dynamic recovery reach a balance state, and the stress–strain curves become constant, until obvious damage occurs inside the solder joints. In the last stage of the thermal fatigue process, the nominal stress decreases due to the interfacial damage, while the real strain increases gradually, making the nominal stress-real strain curve similar to the curve shown in Fig. 5.16b. With the real strain increasing, damage of the solder joint accelerates.

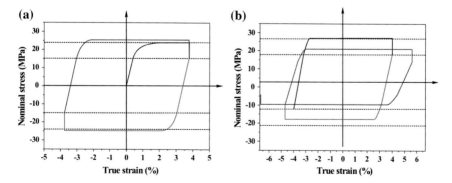

Fig. 5.16 The nominal stress-true strain curves of the solder during the **a** initial cycles and **b** ultimate cycles of the thermal fatigue processes. Reprinted from Ref. [18], Copyright 2013, with permission from Elsevier

For all the thermal fatigued solder joints shown above, there is obvious deformation bands appeared at the surface of the solder, which become wider and deeper with increasing cycles, indicating that dislocation slip is one of the primary deformation mechanisms of the solder. Besides, due to the high homogeneous temperature, significant creep can also occur. During the heating/cooling stage, the strain rate is relatively high, and plastic deformation should be the primary deformation mechanism, while creep is more important during the dwelling stage, inducing the stress relaxation. Since the homologous temperature (T/T_m) is higher than 0.6, dynamic high temperature recovery occurs during the deforming process, the dislike dislocations disappear through destruction. As a result, the strain energy and dislocation density in the solder keep at a low level, and grain subdivision and recrystallization do not occur. At higher strain amplitude, plastic deformation of the solder is more serious, and the deformation bands are deeper. In addition, strain concentration occurs inside the solder around the joint interface, especially at the corner of the solder joint, making plastic deformation the major deformation mechanisms of the solder in these regions.

It has been well accepted that dislocation climb and grain-boundary sliding are two major creep deformation mechanisms of the Sn-based solder alloys [35–38], and their contributions depend on the strain rate, stress level, and temperatures [39]. Dislocation climb is the primary mechanism at higher strain rate, higher stress level, and lower homologous temperature, while grain-boundary sliding is more favored at higher homologous temperature and lower strain rate [2]. Because of its relatively low homologous temperature, dislocation climb may show a greater dominance in creep deformation of the Sn–Ag solders [1]. However, since the strain rate is very low during the thermal fatigue process, grain-boundary sliding is also obvious, as observed in evolution of the grains maps. The inner-grain creep and grain-boundary sliding show good deformation compatibility, thus the cracking at the grain boundaries is slight.

On account of its low melting temperature, the microstructure evolution such as recovery and recrystallization can occur in the Sn–Ag solder at room temperature. At the initial stage of the thermal fatigue process, dislocation slip and multiplication keep developing, which makes the dislocation density increase continually. When the dislocation density increases to a certain value, high temperature recovery promoted by the strain energy occurs rapidly inside the solder [7, 11, 12, 24, 30], i.e., the dislocations disappear through offset of the dislike dislocations, or rearrange into sub-grain boundaries (misorientation angle $\Delta\theta < 2°$) to decrease the dislocation density [40–42]. With increasing thermal fatigue cycles, the sub-grain rotation occurs in the solder, resulting in a continuous increment in their misorientation angle and a transition from sub-grain boundaries to low-angle grain boundaries ($2° < \Delta\theta < 15°$). In some thermal fatigue tests with high strain amplitude, grain subdivision and recrystallization usually occur in the solder. Whereas, due to the low strain amplitude and the long cycle time in this study, the dynamic recovery can keep the strain energy at a low level. Therefore, grain subdivision or recrystallization can hardly occur inside the solder, except for the solder at the strain concentration regions. As the strain amplitude suffered by the solder joints in the real service condition is very low, the thermal fatigue damage behavior in this study should be more similar to the damage of the solder joint in the real service environment.

5.4.3 Influencing Factors on Thermal Fatigue Life

Based on the understandings on the failure mechanisms, some intrinsic and extrinsic factors can affect the thermal fatigue resistance of the solder joints. The temperature/strain amplitude is obviously the major extrinsic influencing factor. At higher temperature/strain amplitude, the deformation and interfacial strain concentration is much serious, and the damage accumulation rate is higher, i.e., the damage is easier to accumulate, inducing higher damage rate and lower life time. Besides, at higher strain amplitude, the strain energy in the solder is higher, indicating that the plastic deformation may contribute higher proportion to the total strain. The thermal fatigue behavior is also related to the strain rate (the temperature changing rate), plastic deformation is more favored at higher strain rate [28], and the recovery is difficult to occur since the time is insufficient, making the solder joint damage at higher rate.

The intrinsic influencing factors include the geometry and microstructure of the solder joints. Since the mechanical property of solder is dominated by its microstructure, in consequence microstructure of the solder has significant influence on the thermal fatigue behaviors. The solders composed by fine grains have higher yield strength and therefore higher resistance to plastic deformation and crack initiation. The interfacial microstructure of solder joints may affect the crack initiation behavior, because the microcracks are formed due to interfacial deformation of the solder. As the strength and interfacial fracture behavior of the solder joint are affected by the IMC thickness [19, 43], the interfacial IMC layer may also affect the thermal fatigue resistance. The geometry of the solder joints are important external factors in deciding the thermal fatigue behavior, because it has significant influence on the strain localization behavior [20], and in turn can affect the crack initiation and propagation process. In addition, the defects in the solder joints are usually the crack initiation location, thus they usually decrease the thermal fatigue resistance.

5.5 Brief Summary

The thermal fatigue behaviors of the Sn–4Ag/Cu solder joints were revealed comprehensively by SEM and EBSD. Based on the experimental results and discussions, the main conclusions can be drawn:

(1) The thermal fatigue processes of the Sn–Ag/Cu solder joints consist of the strain hardening stage and the accelerating fracture stage. During the initial

cycles, the strain hardening keeps developing, and the strength of the solder joint increases a little bit. When the hardening increases to some degree, it balances with dynamic recovery, and the strength keeps stable. The damage at the strain concentration region is serious and can not release through recovery, resulting in microcracks around the joint interface at the corner of the solder joint, then the damage accelerates with increasing cycles.

(2) Thermal fatigue deformation of the solder is the sum of plastic deformation and creep, deformation bands appear on the surface of the solder, but there is little cracking in the solder except for the strain concentration region. Since the dynamic recovery keeps the strain energy in the solder at a low level, grain subdivision, and recrystallization do not occur. Dislocation climb and grain-boundary sliding are the major creep deformation mechanism.

(3) The strain amplitude and geometry of the solder joints are major influencing factors on thermal fatigue behaviors. Interfacial plastic deformation of the solder is much more serious at higher strain amplitude, inducing lower life time. Microstructure of the solder can significantly affect its deformation behaviors, and the interfacial microstructure affects the crack initiation process. The geometry of the solder joints can affect the thermal fatigue damage through influencing the strain concentration.

References

1. Zhang QK, Zhang ZF. In situ observations on creep fatigue fracture behavior of Sn-4Ag/Cu solder joints. Acta Mater. 2011;59:6017–28.
2. Evans JW. A guide to lead-free solders. 1st ed. London: Springer; 2005.
3. Zhang QK, Zhang ZF. In situ tensile creep behaviors of Sn-4Ag/Cu solder joints revealed by electron backscatter diffraction. Scripta Mater. 2012;67:289–92.
4. Guo F, Choi S, Subramanian KN, Bieler TR, Lucas JP, Achari A, et al. Evaluation of creep behavior of near-eutectic Sn-Ag solders containing small amount of alloy additions. Mater Sci Eng A. 2003;351:190–9.
5. Telang AU, Bieler TR, Crimp MA. Grain boundary sliding on near-7 degrees, 14 degrees, and 22 degrees special boundaries during thermornechanical cycling in surface-mount lead-free solder joint specimens. Mater Sci Eng A. 2006;421:22–34.
6. Abd El-Rehim AF. Effect of grain size on the primary and secondary creep behavior of Sn-3 wt.% Bi alloy. J Mater Sci. 2008;43:1444–50.
7. Terashima S, Takahama K, Nozaki M, Tanaka M. Recrystallization of Sn grains due to thermal strain in Sn-1.2Ag-0.5Cu-0.05Ni solder. Mater Trans. 2004;45:1383–90.
8. Telang AU, Bieler TR, Zamiri A, Pourboghrat F. Incremental recrystallization/grain growth driven by elastic strain energy release in a thermomechanically fatigued lead-free solder joint. Acta Mater. 2007;55:2265–77.
9. Li J, Xu H, Mattila TT, Kivilahti JK, Laurila T, Paulasto-Kröckel M. Simulation of dynamic recrystallization in solder interconnections during thermal cycling. Comput Mater Sci. 2010;50:690–7.
10. Choi S, Subramanian KN, Lucas JP, Bieler TR. Thermomechanical fatigue behavior of Sn-Ag solder joints. J Electron Mater. 2000;29:1249–57.

11. Telang AU, Bieler TR, Zamiri A, Pourboghrat F. Incremental recrystallization/grain growth driven by elastic strain energy release in a thermomechanically fatigued lead-free solder joint. Acta Mater. 2007;55:2265–77.

12. Li J, Xu H, Mattila TT, Kivilahti JK, Laurila T, Paulasto-Kröckel M. Simulation of dynamic recrystallization in solder interconnections during thermal cycling. Comput Mater Sci. 2010;50:690–7.

13. Lee JG, Telang A, Subramanian KN, Bieler TR. Modeling thermomechanical fatigue behavior of Sn-Ag solder joints. J Electron Mater. 2002;31:1152–9.

14. Sidhu RS, Chawla N. Thermal fatigue behavior of Sn-Rich (Pb-free) solders. Metall Mater Trans A. 2008;39A:799–810.

15. Choi S, Lee JG, Subramanian KN, Lucas JP, Bieler TR. Microstructural characterization of damage in thermomechanically fatigued Sn-Ag based solder joints. J Electron Mater. 2002;31:292–7.

16. Terashima S, Tanaka M, Tatsumi K. Thermal fatigue properties and grain boundary character distribution in Sn-xAg-0·5Cu (x = 1, 1·2 and 3) lead free solder interconnects. Sci Technol Weld Joining. 2008;31:60–5.

17. Chawla N. Thermomechanical behaviour of environmentally benign Pb-free solders. Int Mater Rev. 2009;54:368–84.

18. Zhang QK, Zhang ZF. Thermal fatigue behaviors of Sn–4Ag/Cu solder joints at low strain amplitude. Mater Sci Eng A. 2013;580:374–84.

19. Dao M, Chollacoop N, Van Vliet KJ, Venkatesh TA, Suresh S. Computational modeling of the forward and reverse problems in instrumented sharp indentation. Acta Mater. 2001;49:3899–918.

20. Deng X, Sidhu RS, Johnson P, Chawla N. Influence of reflow and thermal aging on the shear strength and fracture behavior of Sn-3.5Ag solder/Cu joints. Metall Mater Trans A. 2005;36A:55–64.

21. Shen YL, Chawla N, Ege ES, Deng X. Deformation analysis of lap-shear testing of solder joints. Acta Mater. 2005;53:2633–42.

22. Moy WH, Shen YL. On the failure path in shear-tested solder joints. Microelectron Reliab. 2007;47:1300–5.

23. Matin MA, Vellinga WP, Geers MGD. Thermomechanical fatigue damage evolution in SAC solder joints. Mater Sci Eng A. 2007;445:73–85.

24. Rhee H, Subramanian KN. Effects of prestrain, rate of prestrain, and temperature on the stress-relaxation behavior of eutectic Sn-3.5Ag solder joints. J Electron Mater. 2003;32:1310–6.

25. Lee YH, Lee HT. Shear strength and interfacial microstructure of Sn-Ag-xNi/Cu single shear lap solder joints. Mater Sci Eng A. 2007;444:75–83.

26. Zhao J, Cheng CQ, Qi L, Chi CY. Kinetics of intermetallic compound layers and shear strength in Bi-bearing SnAgCu/Cu soldering couples. J Alloys Compd. 2009;473:382–8.

27. Zhang QK, Zhang ZF. Fracture mechanism and strength-influencing factors of Cu/Sn-4Ag solder joints aged for different times. J Alloy Compd. 2009;485:853–61.

28. Ohguchi KI, Sasaki K, Ishibashi M. A quantitative evaluation of time-independent and time-dependent deformations of lead-free and lead-containing solder alloys. J Electron Mater. 2006;35:132–9.

29. Mavoori H, Chin J, Vayman S, Moran B, Keer L, Fine M. Creep, stress relaxation, and plastic deformation in Sn-Ag and Sn-Zn eutectic solders. J Electron Mater. 1997;26:783–90.

30. Jadhav SG, Bieler TR, Subramanian KN, Lucas JP. Stress relaxation behavior of composite and eutectic Sn-Ag solder joints. J Electron Mater. 2001;30:1197–205.

31. Kerr M, Chawla N. Creep deformation behavior of Sn-3.5Ag solder/Cu couple at small length scales. Acta Mater. 2004;52:4527–35.

32. Shohji I, Yoshida T, Takahashi T, Hioki S. Tensile properties of Sn-Ag based lead-free solders and strain rate sensitivity. Mater Sci Eng A. 2004;366:50–5.

33. Fouassier O, Heintz JM, Chazelas J, Geffroy PM, Silvain JF. Microstructural evolution and mechanical properties of SnAgCu alloys. J Appl Phys. 2006;100:043519.

34. Zhu FL, Zhang HH, Guan RF, Liu S. Effects of temperature and strain rate on mechanical property of Sn96.5Ag3Cu0.5. Microelectron Eng. 2007;84:144–50.
35. McCabe RJ, Fine ME. Creep of tin, Sb-solution-strengthened tin, and Sb Sn–precipitate-strengthened tin. Metall Mater Trans A. 2002;33:1531–9.
36. Haung ML, Wang L, Wu CML. Creep behavior of eutectic Sn-Ag lead-free solder alloy. J Mater Res. 2002;17:2897–903.
37. Sharma P, Dasgupta A. Micro-mechanics of creep-fatigue damage in PB-SN solder due to thermal cycling—part II: mechanistic insights and cyclic durability predictions from monotonic data. J Electron Pack. 2002;124:298–304.
38. Sharma P, Dasgupta A. Micro-mechanics of creep-fatigue damage in Pb-Sn solder due to thermal cycling-part I: formulation. J Electron Pack. 2002;124:292–7.
39. Kashyap BP, Murty GS. Experimental constitutive relations for high-temperature deformation of a Pb-Sn eutectic alloy. Mater Sci Eng. 1981;50:205–13.
40. Wu X, Tao N, Hong Y, Xu B, Lu J, Lu K. Microstructure and evolution of mechanically-induced ultra fine grain in surface layer of AL-alloy subjected to USSP. Acta Mater. 2002;50:2075–84.
41. Liu Q, Jensen DJ, Hansen N. Effect of grain orientation on deformation structure in cold-rolled polycrystalline aluminum. Acta Mater. 1998;46:5819–38.
42. Hansen N, Huang X, Hughes DA. Microstructural evolution and hardening parameters. Mater Sci Eng A. 2001;317:3–11.
43. Deng X, Chawla N, Chawla KK, Koopman M. Deformation behavior of (Cu, Ag)-Sn intermetallics by nanoindentation. Acta Mater. 2004;52:4291–303.

Chapter 6
Conclusions

(1) Fracture behavior of Cu_6Sn_5 IMC at Pb-free solder joint interface

Under indentation loading, cleavage fracture occurs at the foundation or the center portion of the Cu_6Sn_5 grains when the shear stress increases to a certain value. The slender Cu_6Sn_5 grains are more likely to fracture at the foundations, induced by shear stress and the flexural torque, while the podgy grains tend to fracture at the center portion, a result due to the shear stress. The shear fracture strength of the Cu_6Sn_5 grains is about 670 MPa.

After reflowed for a long time, some protrudent Cu_6Sn_5 grains appear at the Sn-4Ag/Cu joint interface, and their shapes can be approximately described with by the revolution body of the parabola. Serious strain concentration occurs around the joint interface during the tensile process. The shear stress applied on the protrudent Cu_6Sn_5 grains at the joint interface reflowed for 8 min are calculated to be about $300 \sim 400$ MPa. At high strain rate, the solder joint is more likely to fracture in the IMC layer.

When the Cu substrate deforms, the dislocations pile-up at the IMC/Cu interface and generate a high cumulative stress ahead the pile-up group after the Cu substrates yield. Fractures inside the IMC layer occur when the cumulative stress reaches the fracture strength of the IMC, the microcracks can propagate to the solder/IMC interface quickly and forms vertical crack under tensile stress. Plastic deformation of the Cu substrates is the sufficient condition for fracture of the IMC layer, and the grain size and yield strength of the Cu substrate has dominative influence on the fracture behaviors.

(2) Tensile-compression fatigue behavior of Pb-free solder joints

Under tensile-compression fatigue loadings, the fatigue life of Sn-4Ag/Cu, Sn-58Bi/Cu, and Sn-37Pb/Cu solder joints decreases exponentially with increasing stress amplitude. Fatigue lives of both the Sn-4Ag/Cu and the Sn-58Bi/Cu joints are

© Springer-Verlag Berlin Heidelberg 2016
Q. Zhang, *Investigations on Microstructure and Mechanical Properties
of the Cu/Pb-free Solder Joint Interfaces*, Springer Theses,
DOI 10.1007/978-3-662-48823-2_6

higher than the Sn-37Pb/Cu solder joints. Due to the deformation mismatch of the solder and substrate, strain concentration occurs at the solder/IMC interface, and fatigue crack initiates along or around the interface. The fatigue crack in the Sn-4Ag/Cu and Sn-37Pb/Cu solder joint propagates inside the solder, close to the joint interface, while the fatigue crack in the Sn-58Bi/Cu solder joint propagates along the solder/IMC interface. When the real stress suffered by the joint interface increases to the interfacial fracture strength, a final fracture similar to the fracture at high strain rate occurs.

The fatigue life consists of crack initiation and propagation cycles. The ductility of solder can significantly affect the fatigue life of solder joint through influencing the crack initiation behavior, crack propagation path, and propagation rate. The thickness of the interfacial IMC layer has little influence on crack initiation and propagation mechanisms, but it can affect the fatigue life by dominating the fracture strength of solder joints. The solder joints aged for different times have similar crack initiation process and initiation cycles, only the different tensile strength leads to different crack propagation cycles and different life cycles.

(3) Creep-fatigue behavior of Pb-free solder joints

The creep-fatigue process of the Sn-4Ag/Cu solder joints consist of a strain hardening stage, a steady deformation stage, and an accelerating fracture stage. During the initial cycles, the strain increases rapidly, until strain hardening reaches a saturated state. After that the strain increases linearly with increasing number of cycles, deformation of the solder continues to develop, and strain concentration occurs around the solder/Cu_6Sn_5 interface and generates initial microcracks. When the microcracks evolve into long cracks, creep-fatigue failure is accelerated and the solder joints fracture along the solder/Cu interface after a few more cycles. The creep-fatigue strain is contributed by plastic deformation and creep of the solder. Grain subdivision occurs in the solder when the plastic strain reaches a threshold, and then grain rotation takes place in the newly formed grains to accommodate further straining. Dislocation climb is the major creep mechanism, and grain boundary sliding can also be promoted after subdivision. The creep-fatigue fracture process of the Sn-58Bi/Cu solder joint can be divided into strain hardening stage, exponential deforming stage, and final fracture stage; fracture occurs inside the solder near the joint interface. The major creep deformation mechanism of the SnBi solder is grain boundary sliding. Plastic deformation concentrates at the grain boundary, while the deformation inside the solder grain is little. Stress amplitude, average stress and holding time are the major external factors influencing the creep-fatigue behavior. The microstructure of the solder can affect its deformation behavior, and the interfacial microstructure dominates crack initiation and final fracture.

(4) Thermal fatigue behavior of Pb-free solder joint

The thermal fatigue processes of the Sn-4Ag/Cu solder joints consist of the strain hardening stage and the accelerating fracture stage. During the initial cycles, the strain hardening keeps developing, and the strength of the solder joint increases a little bit. When the hardening increases to some degree, it balances with dynamic recovery, and the strength keeps stable. The damage at the strain concentration region is serious and cannot release through recovery, resulting in microcracks around the joint interface at the corner of the solder joint; after that the damage accelerates with increasing cycles.

Thermal fatigue deformation of the solder is the sum of plastic deformation and creep; deformation bands appear on the surface of the solder, but there is little cracking in the solder except for the strain concentration region. Since the dynamic recovery keeps the strain energy in the solder at a low level, grain subdivision and recrystallization do not occur. Dislocation climb and grain boundary sliding are the major creep deformation mechanisms. The strain amplitude and geometry of the solder joints are major influencing factors on thermal fatigue behaviors. Interfacial plastic deformation of the solder is much more serious at higher strain amplitudes, inducing lower life time. Microstructure of the solder can significantly affect its deformation behaviors, and the interfacial microstructure affects the crack initiation process. The geometry of the solder joints can affect the thermal fatigue damage by influencing the strain concentration.